Applied Mathematical Sciences
Volume 103

Applied Mathematical Sciences

(continued following index)

Alexandre J. Chorin

Vorticity and Turbulence

With 45 Illustrations

Springer-Verlag
New York Berlin Heidelberg London Paris
Tokyo Hong Kong Barcelona Budapest

Alexandre J. Chorin
Department of Mathematics
University of California
Berkeley, CA 94720
USA

Series Editors

F. John
Courant Institute of
  Mathematical Sciences
New York University
New York, NY 10012
USA

J.E. Marsden
Department of
  Mathematics
University of California
Berkeley, CA 94720
USA

L. Sirovich
Division of
  Applied Mathematics
Brown University
Providence, RI 02912
USA

Mathematical Subject Classification (1991): 76Fxx, 60J65, 62M40

Library of Congress Cataloging-in-Publication Data
Chorin, Alexandre Joel.
  Vorticity and Turbulence / Alexandre Chorin.
    p.   cm. – (Applied mathematical sciences : v. 103)
  Includes bibliographical references and index.
  ISBN 0-387-94197-5 (New York : alk. paper). – ISBN 3-540-94197-5
(Berlin : alk. paper)
  1. Vortex-motion.   2. Turbulence.   3. Fluid mechanics.   I. Title.
  II. Series : Applied mathematical sciences (Springer-Verlag New York
Inc.) : v. 103.
  QC159.C48 1994
  532'.0527–dc20   93-4311

Printed on acid-free paper.

Production managed by Ken Dreyhaupt; manufacturing supervised by Genieve Shaw.
Photocomposed copy prepared from the author's $\mathcal{AMS}$-LaTeX files.
Printed and bound by R.R. Donnelley & Sons, Harrisonburg, VA.
Printed in the United States of America.

9 8 7 6 5 4 3 2 1

ISBN 0-387-94197-5 Springer-Verlag New York Berlin Heidelberg
ISBN 3-540-94197-5 Springer-Verlag Berlin Heidelberg New York

Preface

This book provides an introduction to the theory of turbulence in fluids based on the representation of the flow by means of its vorticity field. It has long been understood that, at least in the case of incompressible flow, the vorticity representation is natural and physically transparent, yet the development of a theory of turbulence in this representation has been slow. The pioneering work of Onsager and of Joyce and Montgomery on the statistical mechanics of two-dimensional vortex systems has only recently been put on a firm mathematical footing, and the three-dimensional theory remains in parts speculative and even controversial.

The first three chapters of the book contain a reasonably standard introduction to homogeneous turbulence (the simplest case); a quick review of fluid mechanics is followed by a summary of the appropriate Fourier theory (more detailed than is customary in fluid mechanics) and by a summary of Kolmogorov's theory of the inertial range, slanted so as to dovetail with later vortex-based arguments. The possibility that the inertial spectrum is an equilibrium spectrum is raised.

The remainder of the book presents the vortex dynamics of turbulence, with as little mathematical and physical baggage as is compatible with clarity. In Chapter 4, the Onsager and Joyce-Montgomery discoveries in the two-dimensional case are presented from a contemporary point of view, and more rigorous recent treatments are briefly surveyed. This is where the peculiarities of vortex statistics, in particular negative (trans-infinite) temperatures, first appear. Chapter 5 summarizes the fractal geometry of vortex stretching, and Chapter 6 provides a brief but self-contained introduction to the tools needed for further analysis, in particular polymer statistics, percolation, and real-space renormalization. In Chapter 7, these tools are used to analyze a simple model of three-dimensional vortex statistics. The Kolmogorov theory is revisited; a rationale is provided for the effectiveness of some large-eddy approximations; and an instructive contrast is drawn between classical and superfluid turbulence.

Some practical information about approximation procedures is provided in the book, as well as tools for assessing the plausibility of approximation schemes. The emphasis, however, is on the understanding of turbulence—its origin, mechanics, spectra, organized structures, energy budget, and renormalization. The physical space methodology is natural, and makes the reasoning particularly straightforward. Open questions are indicated as such throughout the book.

February 1993

Contents

Introduction

This book provides an introduction to turbulence in vortex systems, and to turbulence theory for incompressible flow described in terms of the vorticity field. I hope that by the end of the book the reader will believe that these subjects are identical, and constitute a special case of fairly standard statistical mechanics, with both equilibrium and non-equilibrium aspects. The special properties of fluid turbulence are due to the unusual constraints imposed by the Euler and Navier-Stokes equations, which include topological constraints in the three-dimensional case. The main consequence of these constraints is that turbulent flows typically have negative temperatures. Despite this peculiar feature turbulence fits well within the standard framework of statistical mechanics; in particular, the Kolmogorov exponent appears as a fairly standard critical exponent and large-eddy simulation appears as a renormalization. In the course of the discussion a comparison with certain properties of vortices in superfluids arises in a natural way, and it points out similarities as well as differences between quantum and classical vortices.

The book is rather concise, but I have tried to make it self-sufficient. I assume that the reader is familiar with the bread-and-butter techniques of mathematical physics—Green's functions, Fourier analysis, distributions—and with basic incompressible hydrodynamics. I have provided introductions to those aspects of probability theory—stochastic processes, statistical mechanics, percolation and polymer physics—that are needed in the analysis. I have tried to give enough mathematics to make the physical ideas understandable, without going overboard. Given a choice between a clear heuristic derivation and a much more difficult mathematical one, I have usually chosen the former; for example, I have stressed the original heuristic derivation of the Joyce-Montgomery equation in preference to the more rigorous and more demanding recent versions. The material in this book has been taught in a graduate course at Berkeley with students drawn from mathematics, physics, and engineering, and I hope it is accessible to a first-year graduate student in any one of these fields.

My main goal is to relate turbulence to statistical mechanics, and many interesting issues that do not contribute to this goal have been omitted. For example, there is no discussion of correlations in time nor of the recent developments in turbulent diffusion. There is no mention of turbulent boundary layers. There is no extensive discussion of mathematical issues relating to the Euler and Navier-Stokes equations beyond their conservation and invariance properties. Implied criticism of some recent theories expressed in spectral variables remains for the most part implied.

The book is organized as follows: The first three chapters constitute a fairly standard introduction to turbulence in incompressible flow. Chapter 1 is a quick survey of incompressible hydrodynamics; Chapter 2 uses probability and random fields to define the Fourier transform and the energy, dissipation, and vorticity spectra of homogeneous flow. Chapter 3 contains an account of the Kolmogorov theory and of intermittency. This account departs in several respects from the usual accounts; I believe that the departures are necessary. The next four chapters present the statistical theory of vortex motion. In Chapter 4 the two-dimensional theory is discussed, following the work of Onsager, Joyce and Montgomery, J. Miller, Robert, and others. Negative temperatures make their appearance. Three-dimensional flow differs from two-dimensional flow mainly because it is dominated by vortex stretching. In Chapter 5 the basic facts about vortex stretching and folding are presented. In Chapter 6 the tools needed for a statistical description of the stretching (percolation and polymer statistics in particular) are introduced, and they are then applied in Chapter 7 to vortex statistics in three dimensions. It is reasonable to view the earlier parts of the book as the background material needed for understanding Chapter 7. The applications to the computation of turbulent flows are also discussed in Chapter 7. Open questions remain, and they are presented as such.

The basic premise that the large scales of turbulent flow in three space dimensions are problem-dependent and that the general theory is an analysis of the small scales is explained in Chapter 3. The observation that renormalization in the vortex setting leads to a form of dealiasing and provides a theoretical justification for large-eddy simulation is presented in Chapter 7.

I hesitated a lot before I put the Kolmogorov theory at the beginning of the analysis of turbulence, before the discussion of vortex statistics is even begun. The Kolmogorov theory, for all its brilliant intuition, is imprecise and, I believe, partly misguided. It does however provide a useful framework for later analysis. Its conclusions are revisited in Chapter 7. Much more can be said about the two-dimensional case than I said in Chapter 4; the interested reader is directed to the references. My main interest is in the three-dimensional case, and the most important aspect of Chapter 4 is that it allowed me to introduce negative temperatures in a context where their existence is beyond doubt. The Williams-Shenoy extension of Kosterlitz-Thouless renormalization to three space dimensions is notoriously controversial; its use in the context of classical fluid mechanics should not be, because in this context the primacy of vortex interactions is not in doubt.

My major conclusion is that turbulence can no longer be viewed as incomprehensible. There are many kinds of turbulence and a wide variety of phenomena; e.g., compressible turbulence differs from incompressible turbulence, quantum turbulence differs from classical turbulence, etc. Yet there is a reasonable and consistent general approach to the problem, one that is in harmony with the equations of motion and with results in other areas of statistical physics, and that provides correct information about practical approaches to problem solving. This is all one can expect to have in a problem with so many different aspects. There are many mathematical questions that remain open, and much work remains to be done both in the general theory and in specific problems, yet the overall picture is clear. Maybe the most startling contention in this book is that the inertial range of turbulence can be described in terms of equilibrium statistical mechanics; the cascade picture, so familiar and so nicely celebrated in verse, describes the formation of the inertial range but not situations where that range is already formed.

In many important respects, and certainly in point of view, this book is a second version of my *Lectures on Turbulence Theory* of 1975. The differences are due to the extensive progress made in the last 20 years in understanding turbulent flow.

A comment is needed about notation: I have used the same symbol (μ) for both chemical potential and for Flory exponents, and the same symbol (Z) for both enstrophy and partition function. Which one of the meanings of these symbols is meant should be apparent from the context. These symbols are commonly used in this way, and I have chosen (possibly mistakenly) to risk confusing the reader slightly here rather than have him or her be confused more when he or she turns to other books or papers. The references are collected at the end. In the text, I have mentioned in footnotes some of the references that are most relevant to whatever topic is at hand, giving enough information to enable the reader to find the full citation in the bibliography at the end of the book.

I have greatly benefitted from discussions with Profs. T. Buttke, P. Colella, G. Corcos, J. Goodman, O. Hald, A. Majda, J. Sethian, F. Sherman and A.K. Oppenheim, whom I warmly thank. The responsibility for all errors, however, is mine alone. I would like to thank the Institute for Advanced Study at Princeton for its hospitality in 1991–92, when a first draft of this book was written, and the Basic Energy Sciences program of the U.S. Department of Energy for its support over the past decade.

1
The Equations of Motion

In this chapter we present the Euler and Navier-Stokes equations for a fluid of constant density, in several forms that will be useful later. In particular, the vorticity and vortex magnetization are introduced, and a first discussion of spectral variables is given. Useful results whose proofs are available in elementary textbooks[1] will be merely stated.

1.1. The Euler and Navier-Stokes Equations

We consider a region $\mathcal{D}$, in either two-dimensional or three-dimensional space, that is filled with fluid. In this book, we shall only consider constant density fluids; the density can be taken equal to 1 without loss of generality. A point in $\mathcal{D}$ has coordinates $\mathbf{x} = (x_1, x_2, x_3)$. Vectors will always be denoted by boldface. Let $\mathbf{u}(\mathbf{x}, t) = (u_1, u_2, u_3)$ denote the velocity field of a fluid particle located at $\mathbf{x}$ at time t. In a fluid of constant density the total flow in and out of any subset of $\mathcal{D}$ must add up to zero; elementary vector calculus yields the equation of continuity

$$(1.1) \qquad\qquad \operatorname{div} \mathbf{u} = 0 \ .$$

---

[1] See, e.g., A. Chorin and J. Marsden, 1979, 1990, 1992.

If one takes a particular material point in the fluid, initially (i.e., when $t = 0$) located at $\mathbf{a}$, one will note that it will move in time. The equation of its trajectory is

$$(1.2) \qquad \frac{d\mathbf{x}(t)}{dt} = \mathbf{u}(\mathbf{x}, t) \; ;$$

integrated with the initial condition $\mathbf{x}(0) = \mathbf{a}$, this equation will yield the trajectory of the fluid particle. The map from the initial locations of the fluid particle to their locations at time t is the flow map, ϕ_t (note that the subscript does not denote differentiation). Since the fluid has constant density, a collection of particles that occupies a volume V will move into sites that occupy an equal volume. Thus if $\boldsymbol{\phi}(\mathbf{x}, t)$ denotes the location of the particle that ϕ_t has moved from $\mathbf{a}$ to $\mathbf{x}$, the Jacobian $J = |\partial\phi_i/\partial x_j|$ must equal 1. The equation $J \equiv 1$ is equivalent to the equation of continuity.

Note, for later use, that the equation of continuity can be written in the symbolic form

$$(1.3) \qquad \sum_i \frac{\partial}{\partial x_i} \frac{dx_i}{dt} = 0,$$

where $\mathbf{x}(t) = (x_1(t), x_2(t), x_3(t))$ and $d\mathbf{x}/dt$ is given by equation (1.2).

If $q(\mathbf{x}, t)$ is a function of position and time in the fluid, its derivative with respect to t is

$$\frac{dq}{dt} = \sum_j \frac{\partial q}{\partial x_j} \frac{dx_j}{dt} + \frac{\partial q}{\partial t} = \sum u_j \partial_j q + \frac{\partial q}{\partial t} = \left(\frac{\partial}{\partial t} + \mathbf{u} \cdot \boldsymbol{\nabla} \right) q \; ,$$

where $\boldsymbol{\nabla} = \left(\frac{\partial}{\partial x_1}, \frac{\partial}{\partial x_2}, \frac{\partial}{\partial x_3} \right)$ is the differentiation vector. We shall often write ∂_1 or ∂_{x_1} for $\frac{\partial}{\partial x_1}$. The operator $(\partial_t + \mathbf{u} \cdot \boldsymbol{\nabla})$ will be denoted by D/Dt (even though it is in fact identical to $\frac{d}{dt}$).

Newton's law, force = mass × acceleration, takes for a fluid the form

$$\frac{D\mathbf{u}}{Dt} = -\text{grad } p + \nu \Delta \mathbf{u} + \mathbf{f} \; .$$

The left-hand side is the acceleration, the right-hand side the force. The mass does not appear because the density is 1. The force is the sum of pressure forces, $-\text{grad } p$, viscous friction forces $\nu \Delta \mathbf{u}$ (Δ is the Laplace operator $\sum \partial_j^2$), where ν is the viscosity coefficient, and body forces $\mathbf{f}$ such as gravity. As usual, in each problem one picks a typical velocity U, a typical length L (and thus a time scale $T = L/U$), and refers the variables to them:

$$\mathbf{x}' = \frac{\mathbf{x}}{L} , \qquad \mathbf{u}' = \frac{\mathbf{u}}{U} , \qquad t' = \frac{t}{T} .$$

After dropping the primes, one obtains the dimensionless form of the equations:

(1.4)
$$\frac{D\mathbf{u}}{Dt} = - \text{ grad } p + \frac{1}{R}\Delta\mathbf{u} + \mathbf{f},$$

where $R = \frac{UL}{\nu}$ is the Reynolds number. If $R^{-1} \neq 0$ these equations are known as the Navier-Stokes equations. If $R^{-1} = 0$ they are known as Euler's equations. Flow with $R^{-1} = 0$ will also be called inviscid. On the boundary $\partial\mathcal{D}$ of a bounded domain $\mathcal{D}$ the appropriate boundary conditions are:

(1.5)
$$\begin{aligned} \mathbf{u} &= 0 \quad \text{when } R^{-1} \neq 0 , \\ \mathbf{u} \cdot \mathbf{n} &= 0 \quad \text{when } R^{-1} = 0 \end{aligned}$$

where $\mathbf{n}$ is the normal to $\partial\mathcal{D}$.

If $\mathcal{D}$ is a bounded domain, and $\mathbf{w}$ a sufficiently smooth vector, $\mathbf{w}$ can be written uniquely as a sum of the form

$$\mathbf{w} = \mathbf{u} + \text{ grad } \phi ,$$

where div $\mathbf{u} = 0$, $\mathbf{u} \cdot \mathbf{n} = 0$ on $\partial\mathcal{D}$. The vectors $\mathbf{u}$ and grad ϕ are orthogonal in function space:

$$\int \mathbf{u} \cdot \text{ grad } \phi \, dx = - \int (\text{ div } \mathbf{u})\phi \, dx = 0 .$$

$\mathbf{u}$ can then be viewed as the orthogonal projection of $\mathbf{w}$ on the space of divergence-free vectors tangential to the boundary; $\mathbf{u} = \mathbb{P}\mathbf{w}$. Clearly, $\mathbb{P}$ grad $\phi = 0$ for all ϕ; div $\mathbf{u} = 0$ implies div $\partial_t\mathbf{u} = 0$, thus equation (1.4) can be written

(1.6)
$$\partial_t\mathbf{u} = \mathbb{P}[-(\mathbf{u} \cdot \nabla)\mathbf{u} + R^{-1}\Delta\mathbf{u} + \mathbf{f}].$$

(Note that $\mathbb{P}$ and Δ do not necessarily commute and thus $\mathbb{P}\Delta\mathbf{u} \neq 0$ in general.) This is the projection form of the Navier-Stokes equations. If $R^{-1} = 0$,

$$\partial_t\mathbf{u} = -\mathbb{P}((\mathbf{u} \cdot \nabla)\mathbf{u} - \mathbf{f}) .$$

We shall also be considering in these notes periodic domains, where $\mathbf{u}$ and grad p must be periodic, and infinite domains, where $\mathbf{w}$ must be square integrable and $\mathbf{u}$ must satisfy a decay condition at infinity. The kinetic energy of the flow is

$$E = \tfrac{1}{2} \int \mathbf{u}^2 dx$$

($\mathbf{u}^2$ means $\sum u_i^2$, $dx \equiv dx_1 dx_2 dx_3$). Its rate of change, assuming the external force $\mathbf{f} = 0$, is

$$\frac{dE}{dt} = \int \mathbf{u} \, \frac{D\mathbf{u}}{Dt} \, dx = \int (-\mathbf{u} \cdot \operatorname{grad} p + R^{-1} \mathbf{u} \cdot \Delta \mathbf{u}) dx$$

$$= R^{-1} \int (\mathbf{u} \cdot \Delta \mathbf{u}) dx$$

$$(1.7) \qquad = -R^{-1} \int (\nabla \mathbf{u})^2 dx \;,$$

the integration by parts being valid for any one of the boundary conditions we have considered ($\mathbf{u} \cdot n = 0$, periodic, decay at infinity). As one can readily work out, $(\nabla \mathbf{u})^2 \equiv \sum_{i,j} \left(\frac{\partial u_i}{\partial x_j} \right)^2$. If $R^{-1} = 0$ and $\mathbf{u}$ is smooth enough for the integral to be finite, $E = $ constant. If $R^{-1} \to 0$ and the integral grows more slowly than R, the same conclusion holds.

Note that we do not have an energy equation of the usual type, in which is asserted the conservation of the sum of the kinetic energy and the "internal energy", i.e., the energy associated with the microscopic vibrations of the fluid. The kinetic energy can decay, if $R^{-1} \neq 0$, and it presumably gets transformed into internal energy, which the equations do not take into account. On the other hand, there is no way for internal energy to become converted into kinetic energy since, by definition, an incompressible fluid does not allow for changes in the fluid's specific volume, and an inspection of the thermodynamic formula for the work done by a fluid shows that the work is zero. A change in the classical, molecular temperature of the fluid, the one that can be measured by a thermometer, can have only a limited effect on the motion of an incompressible fluid by changing the viscosity coefficient. (It will be seen below that the modifiers accompanying the word "temperature" in the preceding sentence make sense when a different temperature is introduced.) This effect of the temperature is, in general, small.

In the case of inviscid, smooth flow there is absolutely no coupling between the molecular motion of the fluid and its macroscopic motion as described by our velocity vector $\mathbf{u}$. The molecular structure of the fluid, its microscopic temperature and entropy, have no effect at all on $\mathbf{u}$, and vice versa.

1.2. Vorticity Form of the Equations

Consider the velocity field at two adjacent points at the same time: $\mathbf{u}(\mathbf{x}, t)$, $\mathbf{u}(\mathbf{x}+\mathbf{h}, t)$, and expand $\mathbf{u}(\mathbf{x}+\mathbf{h}, t)$ in powers of $\mathbf{h}$, neglecting terms of $O(h^2)$. The result can be written in the form

$$(1.8) \qquad\qquad \mathbf{u}(\mathbf{x} + \mathbf{h}) = \mathbf{u}(\mathbf{x}) + \tfrac{1}{2}\boldsymbol{\xi} \times \mathbf{h} + D \cdot \mathbf{h} ,$$

where $\boldsymbol{\xi} = \operatorname{curl} \mathbf{u}$ is the vorticity, and $D = \frac{1}{2}(\nabla\mathbf{u} + (\nabla\mathbf{u})^T)$, with $\nabla\mathbf{u}$ the matrix of partial derivatives of $\mathbf{u}$: $(\nabla\mathbf{u})_{ij} = \partial_i u_j$, and $(\nabla\mathbf{u})^T$ its transpose. D is the deformation matrix. Equation (1.8) can be interpreted as stating that locally the most general motion of a fluid is a sum of rigid body translation, rigid body rotation, and deformation. The rotation vector is $\frac{1}{2}\boldsymbol{\xi}$. In an inviscid flow, in which there are no tangential stresses, there is no mechanism for starting or ending rotation, and thus $\boldsymbol{\xi}$ should play a distinguished role.

To find an equation for $\boldsymbol{\xi}$, take the curl of equation (1.4). Some manipulation of vector identities yields

$$(1.9) \qquad\qquad \frac{D\boldsymbol{\xi}}{Dt} = (\boldsymbol{\xi} \cdot \nabla)\mathbf{u} + R^{-1}\Delta\boldsymbol{\xi} .$$

Remember that $\operatorname{div} \mathbf{u} = 0$; $\operatorname{div} \boldsymbol{\xi} = 0$ since $\boldsymbol{\xi}$ is a curl.

In the special case of two-dimensional flow $\boldsymbol{\xi} = (0, 0, \xi)$, and (1.9) becomes

$$\frac{D\xi}{Dt} = R^{-1}\Delta\xi .$$

Consider for a moment inviscid flow, $R^{-1} = 0$. Let C be a smooth closed curve immersed in the fluid, and define the circulation Γ along C as

$$\Gamma_C = \int_C \mathbf{u}(\mathbf{x}, t) \cdot d\mathbf{s} .$$

Let $C_t = \phi_t(C)$ be the image of C under the flow map. The quantity $\Gamma(t) = \int_{C_t} \mathbf{u}(\mathbf{x}, t) \cdot d\mathbf{s}$ is invariant in time (this is the "circulation theorem"). If Σ_t is a surface that spans C_t, $\Gamma(t) = \int_{\Sigma_t} \boldsymbol{\xi} \cdot d\boldsymbol{\Sigma}$; the invariance of Γ thus displays the privileged role of circulation.

A vortex surface, or vortex sheet, is a surface tangent to $\boldsymbol{\xi}$ at each of its points. By the circulation theorem, a vortex sheet remains a vortex sheet as an inviscid flow evolves. A vortex line is an integral line of the vorticity field; it can be viewed as the intersection of two vortex sheets and thus remains a vortex line after mapping by the flow map of an inviscid flow.

Consider a closed pieced of surface B nowhere tangent to $\boldsymbol{\xi}$, and the vortex lines that emerge from it. This is a vortex tube (an essential object in what will follow). The strength or circulation of a vortex tube is the quantity $\Gamma = \int_\Sigma \boldsymbol{\xi} \cdot d\Sigma$, where Σ is a cross-section of the tube; Γ is constant in space (by the divergence theorem), and when $R^{-1} = 0$, it is also a constant in time (by the circulation theorem).

The circulation theorem shows that vorticity cannot be created ab nihilo. This conclusion remains true in the viscous case $R^{-1} \neq 0$. In two space dimensional inviscid flow vorticity is merely transported by the velocity field. In three space dimensions vortex tubes can stretch or become shorter. As their lengths change, their cross-sections change, and as a result of the circulation theorem the magnitude of $\boldsymbol{\xi}$ can change. Indeed, $D\boldsymbol{\xi}/Dt = (\boldsymbol{\xi} \cdot \nabla)\mathbf{u}$ when $R^{-1} = 0$. The right-hand side, the "stretching term", is the rate of change of $\mathbf{u}$ along a vortex line.

If $\boldsymbol{\xi} = 0$ at $t = 0$, it can become non-zero by generation of vorticity at boundaries (the most frequently occurring case) or through the action of external forces.

The relation $\boldsymbol{\xi} = \mathrm{curl}\ \mathbf{u}$ (remembering div $\mathbf{u} = 0$) can be inverted: if $\mathcal{D}$, the domain occupied by the fluid, is simply connected, div $\mathbf{u} = 0$ implies the existence of $\mathbf{A}$ (a vector potential) such that $\mathbf{u} = \mathrm{curl}\ \mathbf{A}$; $\mathbf{A}$ can be chosen so that div $\mathbf{A} = 0$. A simple manipulation of vector identities yields $\Delta\mathbf{A} = -\boldsymbol{\xi}$. Use of the Green's function of the three-dimensional Laplace operator, in the absence of boundaries, yields

$$\mathbf{A} = \frac{1}{4\pi} \int \frac{1}{|\mathbf{x} - \mathbf{x}'|} \boldsymbol{\xi}(\mathbf{x}')d\mathbf{x}'$$

and thus

$$\mathbf{u} = \mathrm{curl}\ \mathbf{A} = -\frac{1}{4\pi} \int \frac{(\mathbf{x} - \mathbf{x}') \times \boldsymbol{\xi}(\mathbf{x}')}{|\mathbf{x} - \mathbf{x}'|^3} d\mathbf{x}'$$

where $\times$ is a cross product. This is the Biot-Savart law. The convolution $f * g$ of two functions f and g is defined as

$$f * g = \int f(\mathbf{x}')g(\mathbf{x} - \mathbf{x}')d\mathbf{x}' \ .$$

The equation for $\mathbf{u}$ can then be written in the form

$$(1.10) \qquad\qquad\qquad \mathbf{u} = K * \boldsymbol{\xi} \ ,$$

where K is the operator $-(4\pi|\mathbf{x}|^3)^{-1}\mathbf{x}\times$. K can also be written in the form

$$K = \frac{1}{4\pi|\mathbf{x}|^3} \begin{pmatrix} 0 & x_3 & -x_2 \\ -x_3 & 0 & x_1 \\ x_2 & -x_1 & 0 \end{pmatrix}.$$

In two space dimensions, $\mathbf{u} = K * \xi$, where

$$(1.11) \qquad K = \frac{1}{2\pi} \begin{pmatrix} -\partial_2 \\ \partial_1 \end{pmatrix} \log |\mathbf{x}| \ .$$

The kinetic energy of a three-dimensional fluid in an unbounded region has already been defined as

$$E = \tfrac{1}{2} \int \mathbf{u}^2 d\mathbf{x} \ .$$

Writing again $\mathbf{u} = \operatorname{curl} \mathbf{A}$, one obtains

$$E = \tfrac{1}{2} \int [(\mathbf{A} \cdot \xi) - \operatorname{div} (\mathbf{u} \times \mathbf{A})] d\mathbf{x} \ .$$

If ξ has compact support, (i.e., if it is zero outside a bounded set), $\mathbf{u}$ decays for large $|\mathbf{x}|$ as $|\mathbf{x}|^{-2}$ [from (1.10) above], and $\mathbf{A}$ decays as $|\mathbf{x}|^{-1}$. Since the area of a sphere is proportional to $|\mathbf{x}|^2$, the divergence term can be transformed into a surface integral that vanishes at infinity; thus

$$(1.12) \qquad E = \frac{1}{2} \int \mathbf{A} \cdot \xi \, d\mathbf{x} = \frac{1}{8\pi} \int d\mathbf{x} \int d\mathbf{x}' \, \frac{\xi(\mathbf{x}) \cdot \xi(\mathbf{x}')}{|\mathbf{x} - \mathbf{x}'|} \ ,$$

where the formula for $\mathbf{A}$ above has been used. Formula (1.12) will be a key formula in the sequel.

The analogous calculation in two-space dimensions is also important but does not end as happily. The boundary term in two dimensions does not decay as $|\mathbf{x}|$ grows, and thus, after similar manipulations, one obtains

$$(1.13) \qquad E = -\frac{1}{4\pi} \iint \xi(\mathbf{x})\xi(\mathbf{x}') \log |\mathbf{x} - \mathbf{x}'| d\mathbf{x} d\mathbf{x}' + B \ ,$$

where B is a (possibly infinite) boundary term. The fact that Green's function in two dimensions is $-\frac{1}{2\pi} \log |\mathbf{x}|$ has been used.

In two space dimensions, the inviscid equations leave invariant all the integrals

$$\mathbf{I}_k = \int \xi^k(\mathbf{x}) d\mathbf{x} \ , \qquad k = 1, 2, 3, \dots \ .$$

In three space dimensions, the "helicity"

$$\mathcal{H} = \int \xi \cdot \mathbf{u} \, d\mathbf{x}$$

is a constant of the motion when $R^{-1} = 0$; the "impulse"

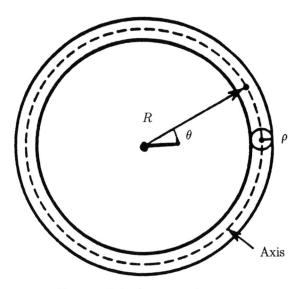

FIGURE 1.1. A vortex ring.

$$\mathbf{I} = \int \mathbf{x} \times \boldsymbol{\xi} \, d\mathbf{x} \,, \qquad \times = \text{ cross product sign },$$

is a constant of the motion for both viscous and inviscid flow. The meaning of $\mathbf{I}$ will be discussed in detail below.

Finally, one can use formula (1.10) to evaluate the velocity of a thin vortex ring of outer radius R and inner radius ρ (Figure 1.1):

$$(1.14) \qquad u = \Gamma(\log(8R/\rho) + C)/4\pi R + O(\rho/R)$$

where $\Gamma = |\mathbf{x}i|\pi\rho^2$ and the constant C depends on the distribution of vorticity within the ring.[2]

1.3. Discrete Vortex Representations

The equations of motion in vorticity form can be discretized so that the result is a set of ordinary differential equations of a particularly simple form. This observation is the origin of vortex-based numerical methods. Our interest in these discrete equations comes from the fact that they have Hamiltonian forms that can be the starting points of statistical mechanical arguments; the conclusions of these arguments can then be generalized to the original equations. In the present section we consider only inviscid flow.

---

[2]H. Lamb, *Hydrodynamics*, Dover, 1932.

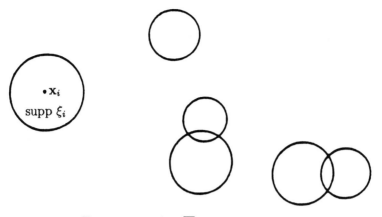

FIGURE 1.2. $\xi = \sum \xi_i$, supp ξ_i small.

The easiest case is that of two-dimensional inviscid unbounded flow. The vorticity field can be written as a sum of functions of small support,

$$\xi \cong \sum_{i=1}^{N} \xi_i(\mathbf{x}) \ .$$

The support of a function ξ_i, supp ξ_i, is the closure of the set of points where it does not vanish; i.e. $\xi_i = 0$ for all $\mathbf{x}$ not in its support (Figure 1.2). A special, useful choice of functions $\xi_i(\mathbf{x})$ is

$$\xi_i(\mathbf{x}) = \Gamma_i \phi_\delta(\mathbf{x} - \mathbf{x}_i) \ , \qquad \phi_\delta = \frac{1}{\delta^2} \phi \left(\frac{\mathbf{x}}{\delta} \right) \ , \qquad \delta \ \text{small} \ ,$$

where ϕ is a smooth function such that $\int \phi \, d\mathbf{x} = 1$, and the Γ_i are coefficients. From (1.11), the velocity field is

$$
\begin{aligned}
\mathbf{u}(\mathbf{x}) \ &= \ \int K(\mathbf{x} - \mathbf{x}')\xi(\mathbf{x}')d\mathbf{x}' \\
&= \ \sum_i \int K(\mathbf{x} - \mathbf{x}')\Gamma_i \phi_\delta(\mathbf{x}' - \mathbf{x}_i)d\mathbf{x}' \\
&= \ \sum K_\delta(\mathbf{x} - \mathbf{x}_i)\Gamma_i \ ,
\end{aligned}
$$

(1.15)

where $K_\delta = K * \phi_\delta$ is a smoothed kernel. It is easy to check that K_δ is smooth when its argument is zero even though K is not, while for large argument $K = K_\delta$.

Consider the motion of the centers of each of the functions of small support, neglecting the deformation of that support by the flow; their velocities are

(1.16)
$$\frac{d\mathbf{x}_i}{dt} = \mathbf{u}(\mathbf{x}_i) = \sum_{j \neq i} K_\delta(\mathbf{x}_i - \mathbf{x}_j)\Gamma_j$$

(The exclusion of $i = j$ is convenient, and at this stage obviously harmless.) The conservation of vorticity expressed by the equation of motion $D\xi/Dt = 0$ yields Γ_j = constant. Equation (1.17) approximates $D\xi/Dt = 0$ as $N \to \infty$. [3]

The amount of writing is least in the special case $\phi_\delta(\mathbf{x}) = \delta(\mathbf{x}) =$ Dirac delta function; let us confine ourselves temporarily to that case. Given the formula (1.11) for K, one can readily see that when $\phi_\delta = \delta$, $K_\delta = K$ and equation (1.16) becomes

(1.17)
$$\frac{dx_{i1}}{dt} = \frac{1}{2\pi} \sum_{j \neq i} \frac{\Gamma_j(x_{j2} - x_{i2})}{r_{ij}^2} ,$$

(1.18)
$$\frac{dx_{i2}}{dt} = -\frac{1}{2\pi} \sum_{j \neq i} \frac{\Gamma_j(x_{j1} - x_{i1})}{r_{ij}^2} ,$$

where $r_{ij} = \sqrt{(x_{i1} - x_{j1})^2 + (x_{i2} - x_{j2})^2} = |\mathbf{x}_i - \mathbf{x}_j|$ and (x_{i1}, x_{i2}) are the components of $\mathbf{x}_i$.

Introduce the Hamiltonian $H = -\frac{1}{4\pi} \sum_i \sum_{j \neq i} \Gamma_i\Gamma_j \log r_{ij}$. Equations (1.17)–(1.18) become

$$\Gamma_i\frac{dx_{1i}}{dt} = \frac{\partial H}{\partial x_{2i}} , \qquad \Gamma_i\frac{dx_{2i}}{dt} = -\frac{\partial H}{\partial x_{1i}} .$$

(no summation over i). Introduce the new variables

$$x'_{1i} = \sqrt{|\Gamma_i|}x_{1i} , \qquad x'_{2i} = \sqrt{|\Gamma_i|}\ \text{sgn}\ (\Gamma_i)x_{2i} ,$$

where $\text{sgn}(\Gamma_i) = 1$ if $\Gamma_i > 0$ and $= -1$ otherwise. Equations (1.17)–(1.18) then become

(1.19)
$$\frac{dx'_{1i}}{dt} = \frac{\partial H}{\partial x'_{2i}} , \qquad \frac{dx'_{2i}}{dt} = -\frac{\partial H}{\partial x'_{1i}} ,$$

i.e., they form a Hamiltonian system. A simple and standard calculation shows that $\frac{dH}{dt} = 0$.

---

[3]See, e.g. A. Chorin, 1972; A. Chorin and P. Bernard, 1973; O. Hald, 1979; J. T. Beale and A. Majda, 1982. For recent reviews, see K. Gustafson and J. Sethian, *Vortex Methods and Vortex Flows*, SIAM Books, 1991, and G. Puckett, 1992.

A similar construction can be carried out for any choice of ϕ_δ; indeed, the possibility of writing the equation in Hamiltonian form follows from the existence of a stream function ψ such that

$$(1.20) \qquad\qquad u_1 = \partial_2 \psi \,, \qquad u_2 = -\partial_1 \psi;$$

the resulting H differs from the one above in that the term $\log r_{ij}$ is smoothed for small r_{ij}. One should notice that the Hamiltonian system (1.19) is rather odd: the variable conjugate to one coordinate of a vortex is the other coordinate of the same vortex [a vorticity field of the form $\Gamma_i \phi_\delta(\mathbf{x} - \mathbf{x}_i)$ will, for simplicity, be called a "vortex"]; in other branches of mechanics one is used to having momenta be conjugate to position variables. Note also that the stream function at x_i is the Hamiltonian at x_i divided by Γ.

One is also used to having the Hamiltonian be equal to the energy of the system, and this is almost the case here. Suppose supp ξ, the support of the vorticity, consists of a collection of N disjoint circles I_i, $i = 1, \ldots, N$ of radii ρ. The expression (1.13) for the energy becomes

$$
\begin{aligned}
E \;=\; & -\frac{1}{4\pi} \sum_i \sum_{i \neq j} \int_{I_i} d\mathbf{x} \int_{I_j} d\mathbf{x}' \log |\mathbf{x} - \mathbf{x}'| \xi(\mathbf{x}) \xi(\mathbf{x}') \\
& -\frac{1}{4\pi} \sum_i \int_{I_i} d\mathbf{x} \int_{I_i} d\mathbf{x}' \log |\mathbf{x} - \mathbf{x}'| \xi(\mathbf{x}) \xi(\mathbf{x}') + B \;.
\end{aligned}
$$

(1.21)

As $\rho \to 0$, with the integral of the vorticity attached to each I_i fixed, the first sum converges to H, the second sum (the "self-energy", which does not affect the motion since it represents the effect of each I_i on itself), becomes a possibly infinite constant, and B is, as before, a possibly infinite constant. If ϕ_δ is smooth and both H and the kinetic energy are finite, H differs from E by a finite constant. Note however that the energy must be positive by definition, while H can be positive or negative depending on the signs of the Γ_i and the values of r_{ij}.

When the flow is confined to a finite domain $\mathcal{D}$ on whose boundary the condition $\mathbf{u} \cdot \mathbf{n} = 0$ is prescribed, the formulas above undergo a slight modification. The Green's function used to relate vorticity to velocity must take into account the boundary; it can be written in the form $G(\mathbf{x}, \mathbf{x}') = -\frac{1}{2\pi} \log |\mathbf{x} - \mathbf{x}'| + \theta(\mathbf{x}, \mathbf{x}')$, where θ is a smooth harmonic function, with $\mathbf{u} = (\partial_2, -\partial_1) \int G(\mathbf{x}, \mathbf{x}') \xi(\mathbf{x}') d\mathbf{x}'$. Smooth terms appear also in the Hamiltonian. Their effects on the analyses below will be small.

In three space dimensions discrete vortex representations are a little harder to set up. One can still write

$$\xi(\mathbf{x}) = \sum_{i=1}^{N} \xi_i(\mathbf{x}) \ ,$$

where the supports of the ξ_i are small. One can try to make these supports spherical, and leave the connectivity constraints to be satisfied weakly. (The connectivity constraints are consequences of the fact that vortex lines in three space dimensions are integral lines of a smooth vector field and are thus connected along substantial distances; furthermore, they satisfy the identity div $\xi = 0$.) This kind of construction is useful numerically[4] but in the analysis below would present problems since the constraints are very important.

One possible choice is to assume that supp ξ_i is a closed vortex tube. The tube has a strength Γ_i, the integral of ξ across a cross-section, that is constant in space and time, and equals the circulation along any contour that surrounds the tube without surrounding any other vortex tubes. To have a finite energy, the tube must have a finite cross-section. Suppose we have N such tubes. At a point $\mathbf{x}$ far from the tubes, the relation (1.10) becomes

$$\mathbf{u}(\mathbf{x}) = -\sum_{i=1}^{N} \frac{\Gamma_i}{4\pi} \int_{i\text{-th tube}} \frac{(\mathbf{x} - \mathbf{x}') \times d\mathbf{s}}{|\mathbf{x} - \mathbf{x}'|^3} \ ,$$

where the integral is along the center-line of the tube and $d\mathbf{s}$ has the obvious meaning of a vector element of length. For $\mathbf{x}$ near $\mathbf{x}'$, the structure of the vorticity field in the tube must be taken into account. One can evaluate $\mathbf{u}(\mathbf{x})$ at points on the tubes and thus advance the tubes. Thin tubes of vorticity will henceforth be referred to as vortex filaments. The stretching of vortex filaments due to the variation of $\mathbf{u}$ along them approximates the stretching term of the equations of motion. Note that in the resulting approximate flow map, the circulation theorem holds.

The energy integral (1.12) becomes

$$E = \frac{1}{8\pi} \sum_i \sum_j \Gamma_i \Gamma_j \int_i \int_j \frac{d\mathbf{s}_i \cdot d\mathbf{s}_j}{|\mathbf{x}(\mathbf{s}_i) - \mathbf{x}(\mathbf{s}_j)|} \ ;$$

with obvious notations. This integral must be smoothed near points where $\mathbf{x}(\mathbf{s}_i) = \mathbf{x}(\mathbf{s}_j)$ by taking into account the finite thickness of the tubes. Much more detail will be given below. For the convergence of this type of approximation as $N \to \infty$, see e.g. Greengard.[5]

---

[4]A. Chorin, 1980; J. T. Beale and A. Majda, *loc. cit*; G. Puckett, *loc. cit*; K. Gustafson and J. Sethian, *loc. cit.*

[5]C. Greengard, 1986. Reviews can be found in A. Leonard, 1985, G. Puckett, 1992.

The disadvantage of this formulation is that it is not obviously Hamiltonian, and that the extended tubes it contains are often hard to manipulate mathematically (these flaws will be remedied in the next section). Its great advantage is that it explicitly takes into account the connectivity of vortex tubes, which will be very important in later developments.

1.4. Magnetization Variables

It is well known that in an appropriate abstract sense Euler's equation in both two and three space dimensions forms a Hamiltonian system.[6] We have exhibited a Hamiltonian structure in two space dimensions through the use of vortices. Hamiltonian formulations are not unique; once one has been found others can be derived from it.

In three space dimensions one specific Hamiltonian formulation that seems to have been discovered independently by several investigators[7] has been shown by Buttke[8] to lead to discrete systems with remarkable properties. The starting point is the introduction of a new variable, $\mathbf{m}$, often referred to rather awkwardly as a magnetization or vortex magnetization (for reasons we shall see), obtained by adding to $\mathbf{u}$ at some point in time an arbitrary gradient:

$$(1.22) \qquad \mathbf{m} = \mathbf{u} + \operatorname{grad} q \qquad \text{at } t = 0 .$$

Obviously at $t = 0$, $\mathbf{u} = \mathbb{P}\mathbf{m}$, with the projection $\mathbb{P}$ defined above. It is not required that div $\mathbf{m} = 0$. We have $\boldsymbol{\xi} = \operatorname{curl} \mathbf{u} = \operatorname{curl} \mathbf{m}$. If one thinks of $\mathbf{u}$ as the vector potential of $\boldsymbol{\xi}$, (1.23) is a gauge transformation of the kind that allows one to add a gradient to the magnetic vector potential in electromagnetic theory without changing the physics. q is not unique, nor is $\mathbf{m}$.

We now proceed in a non-intuitive fashion to find equations for the evolution of $\mathbf{m}$. The end result of our analysis should be heuristically transparent and will justify the effort. We only consider the case of an unbounded domain $\mathcal{D}$.

Consider the following equation for evolving $\mathbf{m} = (m_1, m_2, m_3)$:

$$(1.23) \qquad \frac{Dm_i}{Dt} \equiv \partial_t m_i + u_j \partial_j m_i = -m_j \partial_i u_j$$

(with summation over multiple indices, and $\mathbf{u} = \mathbb{P}\mathbf{m}$). The claim is that the resulting $\mathbf{u}$ is identical to the solution of Euler's equations if $\mathbf{u}(\mathbf{x}, 0) =$

---

[6]See e.g., J. Marsden and A. Weinstein, 1983.

[7]P. Roberts, 1972; V. Oseledets, 1989; A. Rouhi, 1990; T. Buttke, 1992.

[8]T. Buttke, *loc. cit*

$\mathbb{P}\mathbf{m}(\mathbf{x}, 0)$. Equation (1.23) is then the gauge-invariant form of Euler's equations. For the sake of simplicity, we assume there are no external forces.

To check the claim, substitute $\mathbf{m} = \mathbf{u} + \text{grad } q$ into (1.23). After some elementary manipulations, one obtains

(1.24) $\partial_t \mathbf{u} + (\mathbf{u} \cdot \boldsymbol{\nabla})\mathbf{u} = -\text{ grad } (\partial_t q + (\mathbf{u} \cdot \boldsymbol{\nabla})q + \frac{1}{2}|\mathbf{u}|^2)$.

Multiplication by the projection $\mathbb{P}$ yields

$$\partial_t \mathbf{u} + \mathbb{P}((\mathbf{u} \cdot \boldsymbol{\nabla})\mathbf{u}) = 0 ,$$

as promised. Conversely, one can start from equation (1.24) which is equivalent to Euler's equations and obtain (1.23). Note that multiplication of (1.24) by $(I - \mathbb{P})$ (I = identity operator) yields an equation for the evolution of q, which is thus arbitrary only at $t = 0$. It follows from (1.24) that $\mathbf{m}$ and $\mathbf{u}$ always differ by a gradient.

We now have an equation for the evolution of $\mathbf{u}$ plus an initially arbitrary gradient. We shall put this gradient to good use.

Suppose $\boldsymbol{\xi} = \text{curl } \mathbf{u}$ has support within a ball B of finite radius ρ. In three space dimensions, the exterior of a sphere is simply connected, and thus outside B one can write $\mathbf{u} = -\text{grad } \tilde{q}$ for some $\tilde{q}$. Put $q = \tilde{q}$ in (1.22). The resulting $\mathbf{m}$ has support in B. $\mathbf{m}$ can thus be "localized", and this localization persists in time.

Suppose $\boldsymbol{\xi}$ has support in a small sphere B_δ; calculate the resulting $\mathbf{m}$ so that $\mathbf{m}$ also has support in B_δ:

$$\mathbf{m} = \mathbf{M}\phi_\delta(\mathbf{x} - \mathbf{x}_i) ;$$

$\mathbf{x}_i$ is a point in B_δ, $\mathbf{M}$ is a vector coefficient, and $\phi_\delta(\mathbf{x} - \mathbf{x}_i)$ is, as before, a smooth function, with supp ϕ_δ in B_δ, $\int \phi_\delta(\mathbf{x})d\mathbf{x} = 1$. The resulting $\mathbf{u}$ differs from $\mathbf{M}\phi_\delta(x - x_i)$ by a gradient:

$$\mathbf{u} - \mathbf{M}\phi_\delta = K * (\text{ curl } \mathbf{M}\phi_\delta) - \mathbf{M}\phi_\delta = \text{ grad } q ,$$

and thus div $\mathbf{M}\phi_\delta = -\Delta q$, Δ = Laplace operator. Some manipulation of vector identities yields q and then

(1.25) $u_i = M_i\phi_\delta(\mathbf{x} - \mathbf{x}_i) - M_j\partial_j\partial_i\psi_\delta ,$

where $\psi_\delta = \psi_\delta(\mathbf{x} - \mathbf{x}_i)$ satisfies $\Delta\psi_\delta = \phi_\delta$, Δ = Laplace operator.

Consider the magnetization field $\mathbf{m}$ and write it as a sum of N function of small support ("magnets") of the type we have just described:

$$\mathbf{m} = \sum_{i=1}^{N} \mathbf{M}^{(i)}\phi_\delta(\mathbf{x} - \mathbf{x}_i) .$$

The motion of the "centers" $\mathbf{x}_i$ of these functions is of course given by

$$(1.26) \qquad \frac{d\mathbf{x}_i}{dt} = \mathbf{u}(\mathbf{x}_i) = \sum_{j=1}^{N} \mathbf{u}^{(j)}(\mathbf{x}_i) \, ,$$

where $\mathbf{u}^{(j)}$ is the velocity (1.25) due to the j-th "magnet". The coefficients $\mathbf{M}^{(i)}$ are not constants; from equation (1.23) one finds

$$(1.27) \qquad \frac{dM_i^{(k)}}{dt} = -M_j^{(k)} \partial_i u_j(\mathbf{x}_k) \, ,$$

where there is summation over repeated indices and the u_j are the components of $\mathbf{u} = \sum \mathbf{u}^{(k)}$.

One can now check that the flow of these "magnets" is Hamiltonian, with

$$
\begin{aligned}
H \;=\; & \tfrac{1}{2} \sum_j \mathbf{M}^{(j)} \cdot \mathbf{u}(\mathbf{x}_j) \\
\;=\; & \tfrac{1}{2} \sum_{j=1}^{N} \sum_{i=1}^{N} \Big[\mathbf{M}^{(i)} \cdot \mathbf{M}^{(j)} \phi_\delta(\mathbf{x}_i - \mathbf{x}_j) \\
& + (\mathbf{M}^{(i)} \cdot \nabla_i)(\mathbf{M}^{(j)} \cdot \nabla_j) \psi_\delta(\mathbf{x}_i - \mathbf{x}_j) \Big] \, ,
\end{aligned}
$$

(1.28)

where $\nabla_j = (\partial_{x_{j1}}, \partial_{x_{j2}}, \partial_{x_{j3}})$, $\mathbf{x}_j = (x_{j1}, x_{j2}, x_{j3})$, and $\Delta\psi_\delta = \phi_\delta$. If at $t = 0$ the $\mathbf{x}_j$ are distributed so that the sum in (1.28) approximates an integral,

$$H \sim \tfrac{1}{2} \int \mathbf{m} \cdot \mathbf{u} \; dx = \tfrac{1}{2} \int (\mathbf{u} + \operatorname{grad} q) \cdot \mathbf{u} \; dx = \tfrac{1}{2} \int \mathbf{u}^2 \; dx \, ,$$

i.e., the kinetic energy is indeed a Hamiltonian for the flow if one uses the appropriate variables.

One can check that the equations

$$\frac{dx_{jk}}{dt} = \frac{\partial H}{\partial M_k^{(j)}} \, , \qquad \frac{dM_k^{(j)}}{dt} = -\frac{\partial H}{\partial x_{jk}} \, , \qquad [\mathbf{x}_j = (x_{j1}, x_{j2}, x_{j3})] \, ,$$

are exactly equations (1.26) and (1.27).

The "magnets" $\mathbf{M}\phi_\delta$ have a simple interpretation. One can check, by a painful but elementary calculation,[9] that the velocity field (1.25) is the velocity field induced by a small vorticity loop of the form of Figure 1.1, with ρ small, $\mathbf{M}$ perpendicular to the plane of the loop, and $|\mathbf{M}| = \Gamma\pi R^2 =$

---

[9]See, e.g., Jackson, *Classical Electrodynamics*, 1974.

ΓA, where A is the area of the loop and Γ is the circulation $|\boldsymbol{\xi}|\pi\rho^2$. We have thus approximated $\boldsymbol{\xi}$ by a sum of small vortex loops.

There is an analogy between magnetostatics and fluid dynamics, in which the current corresponds to vorticity and the magnetic induction corresponds to velocity; the magnetostatic variables are related by the Biot-Savart law just like the fluid variables. In this analogy, our $\mathbf{m}$ corresponds to the magnetization, hence the name. Since the name is awkward, we shall refer to the loops that the $\mathbf{M}\phi_\delta$ represent as "Buttke loops".

The loop interpretation shows how to convert a vortex representation to a Buttke loop representation. Consider a large vortex loop C of circulation Γ. Construct a surface Σ that spans C. The non-uniqueness of Σ corresponds to the non-uniqueness of q and $\mathbf{m}$. Construct a coordinate system on Σ in terms of some parameters, say s_1, s_2. In each small rectangle $\mathcal{R}$ with vertices (s_1, s_2), $(s_1 + \delta s_1, s_2)$, $(s_1, s_2 + \delta s_2)$, $(s_1 + \delta s_1, s_2 + \delta s_2)$ construct a Buttke loop of strength $\mathbf{M} = \Gamma \delta s_1 \delta s_2$, oriented in an orthogonal direction chosen consistently. The sum of these Buttke loops adds up to the original loop.

The converse problem, how to reconstruct macroscopic vortex loops from an array of Buttke loops, is much harder. How does one decide whether small objects can be viewed as parts of a larger whole? An answer to this kind of question will be given below in Chapter 6, in the context of percolation theory.

Note that for thin closed vortex filaments lying in a plane,

$$\Gamma \int \mathbf{x} \times d\mathbf{s} = 2A\Gamma ,$$

where A is the area surrounded by the filament. Thus

$$\int \mathbf{x} \times \boldsymbol{\xi} \, d\mathbf{x} = 2 \int \mathbf{m} \, d\mathbf{x}$$

and $2\mathbf{m}$ is an impulse density; note that impulse density is thus non-unique. It follows that $\int \mathbf{m} \, d\mathbf{x}$ is a constant of the motion; one can indeed check that $\Sigma \mathbf{M}^{(k)}$ is a constant of the motion for the system (1.26)–(1.27), as is the sum

$$\sum \mathbf{x}_k \times \mathbf{M}^{(k)} ,$$

which is analogous to an angular momentum.

Equation (1.14) in the preceding section shows that the velocity of a small loop varies as a function of the distribution of $\boldsymbol{\xi}$ within the loop. The Hamiltonian formalism must allow for this freedom, as well as for the fact that a Buttke loop has a velocity that is determined by the non-unique choice of ϕ_δ. These facts are reflected in the fact that the addition of a velocity proportional to $\mathbf{M}^{(k)}$ to $d\mathbf{x}_k/dt$ does not destroy the Hamiltonian structure we have described. Indeed, the transformation $\mathbf{u}(\mathbf{x}_k) \to C_k \mathbf{M}^{(k)} + \mathbf{u}(\mathbf{x}_k)$, for each $k = 1, \ldots, N$, C_k arbitrary constants, leaves the system Hamiltonian, with

$$H = \tfrac{1}{2} \sum \mathbf{M}^{(k)} \cdot [\tfrac{1}{2} C_k \mathbf{M}^{(k)} + \mathbf{u}(\mathbf{x}_k)] \ ;$$

thus, equation (1.27) remains valid. For reasonable choices of C_k the convergence of the discrete approximation to the continuum limit is unaffected, since as $N \to \infty$ the "self-velocity" (the effect of the k-th loop on its own velocity) is a shrinking fraction of the total velocity.

Finally, one can readily check that the magnetization form of the Navier-Stokes equations is

$$\partial_t m_i + u_j \partial_j m_i = -m_j \partial_i u_j + R^{-1} \Delta m_i \ , \qquad \mathbf{u} = \mathbb{P}\mathbf{m} \ .$$

1.5. Fourier Representation for Periodic Flow

We now consider a spectral (Fourier) representation for flow in a periodic box with period L: $\mathbf{u}(x_1 + L, x_2, x_3) = \mathbf{u}(x_1, x_2, x_3)$, $\mathbf{u}(x_1, x_2 + L, x_3) = \mathbf{u}(x_1, x_2, x_3)$, etc. Under these conditions, a Fourier series is appropriate. One may wonder why one does not turn immediately to the case of a general domain and a Fourier integral. It will turn out that in those cases where a Fourier integral is useful it must be a Fourier integral of a more general type than we assume the reader is familiar with, and the appropriate discussion must be postponed until the relevant machinery has been introduced. We shall see that the Fourier series is directly useful in several instances, and provides a good preparation for the more general discussion to follow.

Write

$$\mathbf{u}(\mathbf{x}) = \sum_{-\infty}^{+\infty} \hat{\mathbf{u}}_{\mathbf{k}} e_{\mathbf{k}}(\mathbf{x}) \ ,$$

where $e_{\mathbf{k}} = L^{-3/2} \exp(i\mathbf{k} \cdot \mathbf{x})$, $(\mathbf{k} \cdot \mathbf{x}) \equiv \sum_1^3 k_i x_i$, $k_i = \frac{2\pi}{L} \ell_i, \ell_1, \ell_2, \ell_3$ integers, $\hat{\mathbf{u}}_{\mathbf{k}}$ a vector with components $(\hat{\mathbf{u}}_{\mathbf{k}})_1, (\hat{\mathbf{u}}_{\mathbf{k}})_2, (\hat{\mathbf{u}}_{\mathbf{k}})_3$. Since $\mathbf{u}(\mathbf{x})$ is real, $\hat{\mathbf{u}}_{-\mathbf{k}} = \hat{\mathbf{u}}_{\mathbf{k}}^*$, where $*$ denotes a complex conjugate. One can readily check that the $e_{\mathbf{k}}$ form a complete orthonormal periodic set, with period L in each space direction. The equation of continuity for $\hat{\mathbf{u}}_{\mathbf{k}}$ is

(1.29) $\mathbf{k} \cdot \hat{\mathbf{u}}_{\mathbf{k}} = 0 \ ,$

i.e., $\hat{\mathbf{u}}_{\mathbf{k}}$ is orthogonal to $\mathbf{k}$ and thus tangent to a sphere centered at the origin in $\mathbf{k}$-space. The Fourier form of the Navier-Stokes equation is

(1.30) $\partial_t (\hat{\mathbf{u}}_{\mathbf{k}})_\delta = -ik_\gamma \sum_{\mathbf{k}'} (\hat{\mathbf{u}}_{\mathbf{k}'})_\gamma (\hat{\mathbf{u}}_{\mathbf{k}-\mathbf{k}'})_\delta - ik_\delta \hat{p}_{\mathbf{k}} - R^{-1} k^2 (\hat{\mathbf{u}}_{\mathbf{k}})_\delta \ ,$

with $\hat{p}_\mathbf{k}$ the Fourier coefficients of p, the letters γ, δ indices, $\mathbf{k} = (k_1, k_2, k_3)$, $k^2 = \sum_j k_j^2$ and $i = \sqrt{-1}$ (and thus not an index). Elimination of $\hat{p}_\mathbf{k}$ between (1.30) and (1.31) yields

$$(1.31) \qquad \partial_t (\hat{\mathbf{u}}_\mathbf{k})_\alpha = -ik_\delta P_{\alpha\gamma} Q_{\delta\gamma} - R^{-1} k^2 (\hat{\mathbf{u}}_\mathbf{k})_\alpha \ ,$$

where

$$P_{\alpha\gamma} = \delta_{\alpha\gamma} - \frac{k_\alpha k_\gamma}{k^2} \ ,$$

$$\delta_{\alpha\gamma} = \begin{cases} 1 \ , & \alpha = \gamma \\ 0 \ , & \alpha \neq \gamma \end{cases} \ ,$$

$$Q_{\delta\gamma} = \sum_{\mathbf{k}'} (\hat{\mathbf{u}}_{\mathbf{k}'})_\delta (\hat{\mathbf{u}}_{\mathbf{k}-\mathbf{k}'})_\gamma \ ,$$

and, as before, $\hat{\mathbf{u}}_\mathbf{k} = \hat{\mathbf{u}}_{-\mathbf{k}}^*$. The $P_{\alpha\gamma}$ are the components of an operator that projects the vector $k_\delta Q_{\delta\gamma}$ on a plane tangent to a sphere centered at the origin; i.e., the $P_{\alpha\gamma}$ are the components of $\hat{\mathbb{P}}$, the Fourier transform of the projection $\mathbb{P}$ of §(1.1).

The energy of the flow can be written, via the Parseval identity, as

$$\frac{1}{2} \int_{\text{period}} \mathbf{u}^2 \ d\mathbf{x} = \frac{1}{2} \sum_\mathbf{k} |\hat{\mathbf{u}}_\mathbf{k}|^2.$$

Define the m-th "energy shell" as the portion of k-space such that

$$m\Delta < k \leq (m+1)\Delta \ , \qquad k = \sqrt{k_1^2 + k_2^2 + k_3^2} \ ,$$

where $\Delta >> \delta = 2\pi/L =$ spacing between wave numbers in the spectral representation (Figure 1.3). The energy in the m-th energy shell is

$$E_m = \frac{1}{2} \sum_{\text{shell}} |\hat{\mathbf{u}}_\mathbf{k}|^2$$

and thus

$$E = \sum_{m=0}^{\infty} E_m \ .$$

One is tempted to hope that as $L \to \infty$, this last expression tends to something that looks like

$$(1.32) \qquad \text{Energy} = \int_0^\infty E(k) \ dk \ ,$$

for some function $E(k)$. In the present setting this is the case only if the energy remains finite as $L \to \infty$—not a very interesting case, since this implies that the average energy per unit area tends to zero. We shall see

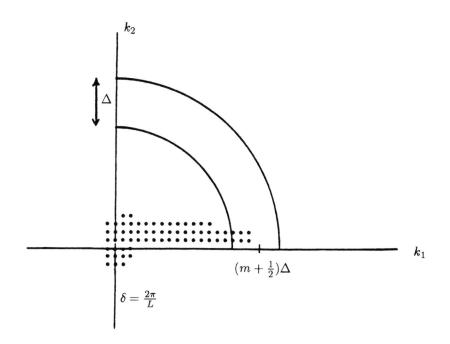

FIGURE 1.3. An energy shell. δ = distance between dots.

however that with the help of probabilistic considerations one can construct an "energy spectrum" $E(k)$ such that the average energy per point is the integral of $E(k)$.

Both the vortex representation and the spectral representation for periodic flows are discrete, in the sense that the number of variables is countable, but not necessarily finite. A vortex representation in a periodic domain can be made finite by picking a finite number of vortices and including the boundary conditions in their interaction, for example, by summing analytically the effects of all the periodic images. A spectral calculation can be made finite by throwing out, in some clever way, all the wave numbers above some large bound $K_{\max}$ (the cleverness is needed to minimize the effect of this truncation on the components that remain). The two types of finite calculations are not equivalent. A finite number of vortices gives rise to velocity and vorticity fields whose Fourier series have an infinite number of non-zero coefficients. A calculation with a finite number of Fourier coefficients, however cleverly designed, cannot be expected to preserve exactly the circulation around closed contours that flow by the approximate flow map. These differences will be highlighted below.

Note that in the discussion of discrete vortex representations we have omitted any mention of what should be done when $R^{-1} \neq 0$, i.e., the discussion was for inviscid flow only. The omission will be repaired in Chapter 2 below. In the spectral case, the viscous term presents no particular problem and was therefore included as a matter of course.

2
Random Flow and Its Spectra

After a short introduction to probability and to random fields, we shall define homogeneous random flow, its energy, vorticity and dissipation spectra, and its Fourier transform.

2.1. Introduction to Probability Theory

We have to formalize the notion of a variable (e.g., a flow) whose values depend on an experiment. We need a set of possible experiments (= "sample space"), a collection of "events" (i.e., experiments with some definite outcome), and some way to assign a probability to these events.

Thus, we pick a space Ω to be our sample space; at this point it is not specified any further.

Let $\mathcal{B}$ be a collection of subsets of Ω (our events), such that whenever $A, B \in \mathcal{B}$, their union $A \cup B$, and their complements $CA, CB \in \mathcal{B}$, and $\emptyset$ = the empty set $\in \mathcal{B}$. (Thus, in particular, $\Omega \in \mathcal{B}$.) In addition, whenever $A_n \in \mathcal{B}$, $n = 1, \ldots, \infty$, $\cup A_n \in \mathcal{B}$. Such a collection is called a σ-algebra. Let μ be a non-negative set function defined on a σ-algebra $\mathcal{B}$ (i.e., a rule which to each member of $\mathcal{B}$ assigns some non-negative number); let μ satisfy the following conditions:

    (i) $\mu(\emptyset) = 0$ (This will become the statement that the probability of an impossible event is zero.)

(ii) $\mu(\bigcup_{n=1}^{\infty} A_n) = \sum_{n=1}^{\infty} \mu(A_n)$ whenever $A_n \cap A_m = \emptyset$, $n \neq m$, and $A_n \in \mathcal{B} \; \forall n$.

μ is called a probability measure if $\mu(\Omega) = 1$ (the probability of something happening is 1); if $\mu(\Omega) = 1$, then the triple $(\Omega, \mathcal{B}, \mu)$ is called a probability space. We shall denote a probability measure by P.

Example: Let Ω be the real line, $\mathcal{B}$ the σ-algebra generated by the half-open sets (i.e., sets of the form $\{x | a < x \leq b\}$, a, b constants), their finite intersections, complements and unions. The sets in this $\mathcal{B}$ are called Borel sets. Given a probability measure P on $(\Omega, \mathcal{B})$, define the function

$$F(x) = P(-\infty < \omega \leq x, \;\; \omega \text{ in } \Omega) .$$

F is called the distribution function of P. We have from the definition:

(i) F is non-decreasing
(ii) F is continuous from the right, i.e.,

$$\lim_{0 < \varepsilon \to 0} F(x + \varepsilon) = F(x) ;$$

(iii) $F(+\infty) = 1$, $F(-\infty) = 0$.

Conversely, $F(x)$, if it satisfies conditions (i)–(iii), defines an appropriate set function on the intervals of the form $a < x \leq b$, and can be extended to their finite intersections, complements and unions.

If F is differentiable, $F'(x)$ is called the probability density of P,

$$F'(x)dx = P(x < \omega \leq x + dx, \;\; \omega \text{ in } \Omega) .$$

Let $(\Omega, \mathcal{B}, P)$ be a probability space. Let $\eta(\omega)$ be a real-valued function defined for $\omega \in \Omega$. Let η satisfy the following condition: For every Borel set S on the real line, the set $\{\omega | \eta(\omega) \in S\} \in \mathcal{B}$ [i.e., one can assign a probability to the event that $\eta(\omega)$ has a numerical value in a certain set]. $\eta(\omega)$ is called a random variable, i.e., a variable whose value depends on an experiment, with probability assigned to the event that it should assume certain values. The integral of $\eta(\omega)$, if it exists, is called the expected value or mean of η and is denoted by $\langle \eta \rangle$:

$$\langle \eta \rangle = \int_{\Omega} \eta(\omega) dP .$$

Note that the random variable η induces a probability measure P_η on the real line through the equation

$$P_\eta(S) = P(\{\omega | \eta(\omega) \in S\}) , \;\;\; S = \text{ Borel set } .$$

P_η is the distribution of η. The function $F_\eta(x) = P(\{\omega|\eta(\omega) \le x\})$ is the distribution function of η; it is identical to the distribution function of P_η. F_η', if it exists, is the probability density of η, and we have

$$\langle \eta \rangle = \int_{\text{real line}} x \, dP_\eta(x) = \int_{-\infty}^{+\infty} x \, dF_\eta(x) = \int_{-\infty}^{+\infty} x F_\eta'(x) \, dx$$

if $F_\eta' = f_\eta$ exists.

A set function μ which satisfies all the conditions above except $\mu(\Omega) = 1$ is a measure (but not a probability measure). An example is the Lebesgue measure on the line, obtained by extension of the usual notion of length.

We shall often use the shorthand $P(\eta \le x)$ for $P\{\omega|\eta(\omega) \le x\}$.

Let $A_1, \ldots, A_n$ be events; i.e., subsets of Ω belonging to $\mathcal{B}$. The events are called independent if for any subcollection $A_{i_1}, A_{i_3}, \ldots, A_{i_q}$,

$$P(A_{i_1} \cap A_{i_2} \cap \cdots \cap A_{i_q}) = P(A_{i_1})P(A_{i_2}) \cdots P(A_{i_q})$$

(i.e., the probability of all of them occurring at once is the product of the probabilities that each one of them should occur). An infinite set of events is said to consist of independent events if any finite subset consists of independent events. The random variables $\eta_1, \ldots, \eta_n$ are independent if the events

$$A_i = \{\omega|\eta_i(\omega) \in S_i\} \ , \quad S_i \text{ a Borel set },$$

are independent.

Let η_1, η_2 be two random variables. The random variable (η_1, η_2) is a vector-valued random variable; it induces a probability measure $P_{\eta_1 \eta_2}$ on the plane, which is called the joint distribution of η_1 and η_2. The function

$$F_{\eta_1 \eta_2} = P(\{\omega|\eta_1(\omega) \le x \ \text{ and } \ \eta_2(\omega) \le y\}$$

is their distribution function. If η_1, η_2 are independent,

$$P_{\eta_1 \eta_2}(S_1 \times S_2) = P_{\eta_1}(S_1)P_{\eta_2}(S_2) \ ,$$

where S_1, S_2 are Borel sets on the line, and $S_1 \times S_2$ is the corresponding "rectangle". Furthermore,

$$F_{\eta_1 \eta_2}(x, y) = F_{\eta_1}(x)F_{\eta_2}(y) \ .$$

Conversely, if the last equation is satisfied, η_1 and η_2 are independent. Furthermore, if η_1 and η_2 are independent,

$$\langle \eta_1 \eta_2 \rangle = \langle \eta_1 \rangle \langle \eta_2 \rangle \ .$$

Let η be a random variable. The numbers

$$(2.1) \qquad E[\eta^n] \equiv \int \eta^n dP = \int x^n dF_\eta(x) = \int x^n f_\eta(x) dx$$

if $F'_\eta = f_\eta$ exists, are called the moments of η. The moments of $\eta - \langle f \rangle = \eta_c$ are called the centered moments of f. In particular,

$$\langle \eta_c^2 \rangle = \text{variance of } \eta = \text{Var}(\eta) ,$$

$$\sqrt{\langle \eta_c^2 \rangle} = \text{standard deviation of } \eta .$$

Let η be a random variable, and G an increasing, non-negative function defined on the range of η; i.e., G is defined for all values which η can assume. Let $P(\eta \geq a)$ be the probability that $\eta \geq a, a$ constant; $P(\eta \geq a) \equiv P(\{\omega | \eta(\omega) \geq a\})$. Assume $G(a) \neq 0$. We have

$$\langle G(\eta) \rangle = \int_{-\infty}^{+\infty} G(\eta) dF_\eta \geq \int_a^\infty G(\eta) dF_\eta$$

$$(2.2) \qquad\qquad \geq G(a) \int_a^\infty dF_\eta = G(a) P(\eta \geq a) .$$

Thus

$$P(\eta \geq a) \leq \frac{\langle G(\eta) \rangle}{G(a)} .$$

In particular, let $G(\eta) = \eta^2$, and $\eta = |g - \langle g \rangle|$ where g is a random variable; we find

$$P(|g - E[g]| \geq a) \leq \frac{\text{var}(g)}{a^2}$$

where var(g) is the variance of g.

Let $\eta_1, \eta_2, \ldots, \eta_n$ be independent random variables, each with the same distribution function and same finite mean m and variance σ^2. Let $\eta = \sum_{i=1}^n \eta_i$. The mean of η is nm, its variance is $n\sigma^2$. We have

$$P\left(\left| \frac{\sum_{i=1}^n \eta_i}{n} - \langle \eta_1 \rangle \right| \geq \varepsilon \right) \leq \frac{\sigma^2}{n\varepsilon^2} , \qquad \varepsilon > 0 ;$$

by the formula above. Thus,

$$\lim_{n \to \infty} P\left(\left| \frac{\sum_{i=1}^n \eta_i}{n} - \langle \eta_1 \rangle \right| \geq \varepsilon \right) = 0 .$$

This agrees with our intuitive idea of "mean" or "average". For later use we can rewrite the formula above in the form

$$P\left(\left|\frac{\Sigma\eta_i}{n} - m\right| \geq \frac{k\sigma}{\sqrt{n}}\right) \leq \frac{1}{k^2}$$

where k is a constant and we wrote $\varepsilon = k\sigma/\sqrt{n}$. For $n = 1$, $P(|\eta - m| \geq k\sigma) \leq 1/k^2$; a variable is not likely to differ from its mean by more than a few standard deviations. The standard deviation gives a rough estimate of the magnitude of $(\eta - \langle\eta\rangle)$.

For later use we wish to single out a particularly important class of random variables. Define the function

$$G_{\sigma,m}(x) = \frac{1}{\sqrt{2\pi\sigma^2}} \int_{-\infty}^{x} e^{-(u-m)^2/2\sigma^2} du \ .$$

A random variable η which admits $G_{\sigma,m}$ as its distribution function is called Gaussian. It is easy to verify that

$$\langle\eta\rangle = m$$

$$\text{var}(\eta) = \langle(\eta - m)^2\rangle = \sigma^2 \ .$$

The importance of Gaussian variables is largely due to the following theorem (the central limit theorem):

Central Limit Theorem *Let $\eta_1, \ldots, \eta_n$ be independent random variables with common distribution, mean m and variance $\sigma^2 < +\infty$. Then*

$$P\left(\frac{\sum_{i=1}^{n}\eta_i - nm}{\sqrt{n}} \leq x\right) \longrightarrow G_{\sigma,0}(x) \ .$$

We are interested in vector-valued Gaussian variables, i.e., in variables $\eta = (\eta_1, \ldots, \eta_n)$ which are, in a sense now to be defined, Gaussian. The joint distribution of two variables has been defined above, and clearly generalizes to n variables. Let us denote the joint distribution function of n variables $\eta_1, \ldots, \eta_n$ by

$$F(x_1, x_2, \ldots, x_n) = P_{\eta_1, \ldots, \eta_n}(\{\omega|\eta_1(\omega) \leq x_1, \ldots, \eta_n(\omega) \leq x_n\}) \ .$$

If F is differentiable, then

$$F(x_1, \ldots, x_n) = \int_{-\infty}^{x_1} \cdots \int_{-\infty}^{x_n} F'(x_1, \ldots, x_n)dx_1, \ldots dx_n \ ,$$

$$F' = \frac{\partial^n}{\partial x_1 \ldots \partial x_n} F \ .$$

A Gaussian random vector $\boldsymbol{\eta} = (\eta_1, \ldots, \eta_n)$ is a vector such that F' exists and has the form

$$F'(x_1, \ldots, x_n) = \frac{1}{(2\pi)^n |\Lambda|} \exp(-\tfrac{1}{2}(\Lambda^{-1}(\mathbf{x} - \mathbf{m}), (\mathbf{x} - \mathbf{m}))$$

where $\mathbf{m} = (m_1, \ldots, m_n)$, $m_i = \langle \eta_i \rangle$, Λ is a positive definite real matrix, and $|\Lambda|$ is the determinant of Λ; this is an obvious generalization of the single-variable case.

The correlation matrix of a real random vector is the matrix Q with entries

$$q_{ij} = \langle (\eta_i - \langle \eta_i \rangle)(\eta_j - \langle \eta_j \rangle) \rangle \, .$$

For a Gaussian random vector, $Q = \Lambda$.

Two random variables η_1 and η_2 are said to be orthogonal (or uncorrelated) if $\langle \eta_1 \eta_2 \rangle = 0$. If η_1 and η_2 are independent, then $\eta_1 - \langle \eta_1 \rangle, \eta_2 - \langle \eta_2 \rangle$ are orthogonal. The converse is of course in general false. However, if η_1, η_2 are Gaussian, and $q_{12} = 0$, then η_1, η_2 are independent.

2.2. Random Fields

Let $\mathcal{D}$ be a region in which a random flow occurs,

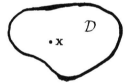

i.e., for each $\mathbf{x} \in \mathcal{D}$, the velocity $\mathbf{u}(\mathbf{x})$ is a random variable; $\mathbf{u}(\mathbf{x}) = \mathbf{u}(\mathbf{x}, \omega)$, $\omega \in \Omega$, with $(\Omega, \mathcal{B}, P)$ a probability space. The knowledge of the distribution of $\mathbf{u}(\mathbf{x})$ for each $\mathbf{x}$ does not provide much information about the flow. For example, assume that $\mathbf{u}(\mathbf{x}, \omega)$ is a Gaussian variable for each $\mathbf{x}$. This statement does not allow one to distinguish between a flow of the form

$$\mathbf{u} = \eta \times \text{constant},$$

where η is a Gaussian random variable, and a flow in which the variables $\mathbf{u}(\mathbf{x}_1, \omega), \mathbf{u}(\mathbf{x}_2, \omega)$, $x_1 \neq x_2$ are independent. The possible samples of the first random flow are very smooth, while those of the second are very wild. (A sample is what one gets for a specific value of ω.) What one needs is information about the relationship between the variables $\mathbf{u}(\mathbf{x}, \omega)$ for the several values of $\mathbf{x}$, so that some global idea about the individual flows may be obtained. Thus, we shall say that a random field is defined if for any finite number n of points $\mathbf{x}_1, \mathbf{x}_2, \ldots, \mathbf{x}_n$, the joint distribution function of the random variables $\mathbf{u}(\mathbf{x}_1), \ldots, \mathbf{u}(\mathbf{x}_n)$ is given.

Let $F_{\mathbf{x}_1, \ldots, \mathbf{x}_n}(\mathbf{y}_1, \ldots \mathbf{y}_n)$ be the joint distribution function of $\mathbf{u}(\mathbf{x}_1, \ldots, \mathbf{x}_m)$;

$$F_{\mathbf{x}_1, \mathbf{x}_2 \cdots \mathbf{x}_n} = P\{\omega | u_1(\mathbf{x}_1) \le y_{11}, u_2(\mathbf{x}_1) \le y_{12}, \ldots, u_m(\mathbf{x}_n) \le y_{nm}\} ;$$

where m = number of dimensions of the space in which the flow occurs, and y_{jp} is the p-th component of the vector $\mathbf{y}_j$. For simplicity, we shall confine ourselves for a while to the scalar case, in which

$$F_{x_1 \cdots x_n} = P\{\omega | u(x_1) \le y_1, \ldots, u(x_n) \le y_n\} .$$

[The existence of the underlying space $(\Omega, \mathcal{B}, \mu)$ is assumed without comment.] The family of functions $F_{x_1 \cdots x_n}(y_1 \cdots y_n)$ must be distribution functions of vector random variables and satisfy the obvious self-consistency requirements:

(i) $F_{x_1 \cdots x_\ell x_{\ell+1} \cdots x_n}(y_1, \ldots, y_\ell, +\infty, +\infty, +\infty \cdots) = F_{x_1 \cdots x_\ell}(y_1 \ldots y_\ell);$
  and

(ii) $F_{x_{i_1} x_{i_2} \ldots x_{i_n}}(y_{i_1} \ldots y_{i_n}) = F_{x_1 \ldots x_n}(y_1 \ldots y_n)$

where $i_1 \ i_2 \ i_3 \ \cdots \ i_n$ is an arbitrary permutation of $1 \ 2 \ 3 \cdots n$.

A random flow will thus be considered defined if the functions $F_{x_1 \cdots x_n}(y_1 \cdots y_n)$ satisfying conditions (i) and (ii) above are given.

The moment functions $m_{j_1 \cdots j_n}(x_1, \ldots, x_n)$ of a random field u are defined by

$$m_{j_1 \cdots j_n} = \langle (u(x_1))^{j_1} \cdots (u(x_n))^{j_n} \rangle ,$$

where $j_1 \cdots j_n$ are nonnegative integers. The mean of $u(x)$ is $\langle u(x) \rangle = m_1(x)$; the correlation function is defined by

$$\mathcal{R}(x_1, x_2) = \langle (u(x_1) - m(x_1))(u(x_2) - m(x_2)) \rangle .$$

In particular, $\mathcal{R}(x, x) = \sigma^2(x)$ = variance of the random variable $u(x)$. If u is complex, we define

$$\mathcal{R}(x_1, x_2) = \langle (u(x_1) - m(x_1))(u(x_2) - m(x_2))^* \rangle ,$$

where the asterisk denotes a complex conjugate.

Some important properties of $\mathcal{R}(x_1, x_2)$ are:

(i) $\mathcal{R}(x_1, x_2) = \mathcal{R}(x_2, x_1)^*$,
(ii) $\mathcal{R}(x_1, x_1) \ge 0$,
(iii) $|\mathcal{R}(x_1, x_2)|^2 \le \mathcal{R}(x_1, x_1)\mathcal{R}(x_2, x_2)$,
(iv) For every n, any $x_1, \ldots, x_n$, and any complex numbers $z_1, \ldots, z_n$,

$$\sum_{j=1}^{n} \sum_{k=1}^{n} \mathcal{R}(x_j, x_k) z_j z_k^* \ge 0 .$$

Conditions (i), (ii) and (iii) are obvious; (iv) follows from

$$\sum_j \sum_k \mathcal{R}(x_j, x_k) z_j z_k^* = \left\langle \sum_j \sum_k (u(x_j) - m(x_j))(u(x_k) - m(x_k))^* z_j z_k^* \right\rangle$$

$$(2.3) \qquad\qquad = \left\langle \left| \sum_j (u(x_j) - m(x_j)) z_j \right|^2 \right\rangle \geq 0 \ .$$

We shall study below the problem of finding what all turbulent flows have in common; it is thus attractive to consider a reasonably simple class of flows and try to analyze its properties; if any general conclusions emerge, they can then be taken to more general cases and their validity tested. An appropriate class of flows is the class of (spatially) homogeneous flows, i.e., flows which fill out all of space and whose statistical properties are independent of the particular point in space one considers. Difficulties with boundary conditions (and boundary layers) are thus bypassed. A discussion of the reasons why a flow can be viewed as random will appear in the next section.

A flow field u is said to be homogeneous if for any n, any $x_1 \cdots x_n$ and any X, the joint distribution of $u(x_1), \ldots, u(x_n)$ is identical to the joint distribution of $u(x_1 + X), \ldots, u(x_n + X)$. In particular, for a homogeneous flow field

$$m_1(x) = m_1 = \text{ constant ;}$$
$$\mathcal{R}(x_1, x_2) = \mathcal{R}(x_1 - x_2, 0) = \mathcal{R}(x_1 - x_2) \ .$$

The properties (i), (ii), (iii) and (iv) of $\mathcal{R}(x_1, x_2)$ become:

  (i) $\mathcal{R}(x) = \mathcal{R}(-x)^*$
  (ii) $\mathcal{R}(0) \geq 0$
  (iii) $|\mathcal{R}(x)| \leq \mathcal{R}(0)$
  (iv) For any $x_1 \cdots x_n$ and $z_1 \cdots z_n$,

$$\sum_j \sum_k \mathcal{R}(x_j - x_k) z_j z_k^* \geq 0 \ .$$

A random flow field is said to be Gaussian if all its finite dimensional distributions are Gaussian.

In order to analyze homogeneous flow fields, we need a representation for the function $\mathcal{R}(x)$, which we shall henceforth call the correlation function. (This terminology differs from the usual statistics terminology and is peculiar to turbulence theory.) We write things in a scalar form for simplicity; the vector case will be exhibited later.

To motivate what follows, we consider an example. Let $u(x)$ be a random flow field. $u^2(x)$ is the "energy" at the point x of the realization $u(x)$, i.e., of $u(x, \omega)$ for the appropriate ω. The integral

$$\int_{-X}^{X} u^2(x) dx$$

is the energy of $u(x)$ in the interval $-X \le x \le X$. The quantity

$$\lim_{X \to \infty} \frac{1}{2X} \int_{-X}^{X} u^2(x) dx \; ,$$

if it exists, will be called the spatial mean energy of $u(x) = u(x, \omega)$. For the sake of exposition, we shall first take the spatial mean of u^2 and then its true mean. Taking the spatial mean first should not be too harmful; it will never be done again after this section.

Consider the field

$$u(x) = \sum_{k=1}^{N} a_k \cos(n_k x + \alpha_k) \; ,$$

where the n_k are given numbers and the a_k, α_k are random variables. $u(x)$ is a random field, fully determined by the joint distribution of the a_k and α_k. The set of frequencies n_k is called the spectrum of $u(x)$. To simplify the analysis, consider the related complex random field

$$u(x) = \sum c_k e^{i n_k x}$$

where $c_k = a_k + i b_k$, and a_k, b_k are real independent random variables. The mean energy of this random field is, if the several means are interchangeable,

$$\left\langle \lim_{X \to \infty} \frac{1}{2X} \int_{-X}^{X} u^2(x) dx \right\rangle = \left\langle \lim \frac{1}{2X} \int_{-X}^{X} \sum_k \sum_r c_k c_r^* e^{ix(n_k - n_r)} dx \right\rangle$$

$$= \lim \left\{ \sum \langle |c_k|^2 \rangle \right.$$

$$\left. + \left\langle \sum_{\substack{k,r \\ k \ne r}} c_k c_r^* \frac{\sin X(n_k - n_r)}{X(n_k - n_r)} \right\rangle \right\}$$

(2.4)
$$= \sum_k \langle |c_k|^2 \rangle \; .$$

The correlation function of $u(x)$ is evaluated similarly:

$$\mathcal{R}(x_1, x_2) = \sum_{k=1}^{N} \langle |c_k|^2 \rangle e^{in_k(x_1 - x_2)} .$$

We also have $m_1(x) = \langle u(x) \rangle = c_0$. Thus $u(x)$ is homogeneous, at least as far as the second moments are concerned—enough for our purposes.

Define the spectral distribution function $F(k)$ by

$$F(k) = \sum_{n_\ell < k} \langle |c_\ell^2| \rangle.$$

(Note the letter F has been used previously for a different purpose; the meaning should be evident from the context. Also, the index has been renamed.) $F(k)$ characterizes the mean power per harmonic component of u. We have

$$\mathcal{R}(x) = \int_{-\infty}^{\infty} e^{ixk} dF(k) .$$

Note that $F(k)$ completely determines, and is completely determined by, the absolute values of the coefficients c_ℓ. However, it is extremely important to notice that all information about the phases of the components c_ℓ, i.e., about ϕ_ℓ in the formula $c_\ell = |c_\ell| \exp(i\phi_\ell)$ is completely lost when $F(k)$ is constructed. For example, there is no way to distinguish, given $F(k)$, between the case where c_ℓ is real for all ℓ and the case where ϕ_ℓ is random.

Let us now proceed to state the more general case and return to the usual definition of an expected value. We have made plausible the following theorem:

(Khinchin): *For a function $\mathcal{R}(x)$, $-\infty < x < +\infty$, to be the correlation function of a field which has translation invariant means and correlation functions and also satisfies the condition*

$$\langle |u(x+h) - u(x)|^2 \rangle \to 0 \quad \text{as} \quad h \to 0 ,$$

it is necessary and sufficient that it have a representation of the form

$$\mathcal{R}(x) = \int_{-\infty}^{+\infty} \exp(ixk) dF(k) .$$

where $F(k)$ is a non-decreasing function of k.

This is a consequence of a classical theorem regarding positive definite functions [e.g., satisfying condition (iv) above].

$F(k)$ is called the spectral distribution function of u. If $F(k)$ is differentiable, $F'(k) = \phi(k)$, and the function ϕ is called the spectral density of u.

Given $F(k)$ [or $\phi(k)$], $\mathcal{R}(x)$ can be reconstructed with the help of Fourier integrals.

More generally, let $\mathbf{u}(\mathbf{x})$ be a vector-valued random field in a multi-dimensional space. Let $\mathbf{u}(\mathbf{x})$ be stationary in the sense that $\langle \mathbf{u}(\mathbf{x}) \rangle$ is a vector independent of x, and the functions $\mathcal{R}_{ij}(\mathbf{r}) = \langle u_i(\mathbf{x}) u_j(\mathbf{x} + \mathbf{r}) \rangle$ are independent of $\mathbf{x}$ and finite (u_i are the components of $\mathbf{u}$). Furthermore, let

$$\langle \| \mathbf{u}(\mathbf{x} + \mathbf{h}) - \mathbf{u}(\mathbf{x}) \|^2 \rangle \to 0 \quad \text{as} \quad \| \mathbf{h} \| \to 0 \ ,$$

where $\| \ \|$ is a vector norm. Then

$$\mathcal{R}_{ij}(\mathbf{r}) = \int \exp(i\mathbf{k} \cdot \mathbf{r}) dF_{ij}(\mathbf{k}),$$

where F_{ij} is such that the matrix with entries $dF_{ij}(\mathbf{k})$ is nonnegative definite, and

$$\sum_i F_{ii}(+\infty) - \sum F_{ii}(-\infty) \ , \quad \infty \equiv (\infty, \infty, \infty, \ldots) \ ,$$

is finite.

Conversely, any $\mathcal{R}_{ij}(\mathbf{r})$ with such a representation is the correlation tensor of a field with the properties above.

We shall henceforth almost always make the assumption that the $F_{ij}(k)$ are differentiable when $\mathbf{u}$ is a velocity field, i.e.,

$$dF_{ij}(\mathbf{k}) = \Phi_{ij}(\mathbf{k}) d\mathbf{k} \ .$$

The reasons for this assumption are mainly (a) that it is convenient, and (b) that it is in agreement with both experiment and our intuitive ideas about the processes by which turbulent flow is generated. The "proofs" contained in the literature contain assumptions which are no more justifiable than the bald statement above. Note that if

$$\langle u_i(\mathbf{x}) u_j(\mathbf{x} + \mathbf{r}) \rangle = \mathcal{R}_{ij}(\mathbf{r}) = \int \exp(i\mathbf{k} \cdot \mathbf{r}) \Phi_{ij}(\mathbf{k}) d\mathbf{k} \ ,$$

then,

$$\tfrac{1}{2} \langle u_i(\mathbf{x}) u_i(\mathbf{x}) \rangle = \text{ mean energy at a point } = \tfrac{1}{2} \int \Phi_{ii}(\mathbf{k}) d\mathbf{k} \ ;$$

thus $\tfrac{1}{2} \Phi_{ii}$ is the density, in wave number space, of the contributions to the mean energy, i.e., the energy density per wave number.

We define

$$E(k) = \tfrac{1}{2} \int_{|\mathbf{k}|=k} \Phi_{ii}(\mathbf{k}) d\mathbf{k}$$

to be the energy spectrum of $\mathbf{u}$; $E(k)$ is the integral of $\Phi_{ii}(\mathbf{k})$ over the sphere of radius $k = |\mathbf{k}|$; it is the contribution to the energy of those wave numbers whose absolute value equals k. Clearly,

$$\tfrac{1}{2}\langle u_i(\mathbf{x})u_i(\mathbf{x})\rangle = \tfrac{1}{2}\langle \mathbf{u}(\mathbf{x})\cdot\mathbf{u}(\mathbf{x})\rangle = \text{ mean energy at a point } = \int_0^\infty E(k)dk \ .$$

[See equation (1.32) in Chapter 1.]

2.3. Random Solutions of the Navier-Stokes Equations

We shall now consider random fields $\mathbf{u}(\mathbf{x},\omega)$ which, for each ω (i.e., for each experiment that produces them), satisfy the Navier-Stokes equations. $\mathbf{u}$ depends also on the time t; we shall usually not exhibit this dependence explicitly.

There is an interesting question of principle that must be briefly discussed: why does it make sense to view solutions of the deterministic Navier-Stokes equations as being random? It is an experimental fact that the flow one obtains in the laboratory at a given time is a function of the experiment. The reason must be that the flow described by the Navier-Stokes equation for large R is chaotic[1]; microscopic perturbations, even on a molecular scale, are amplified to macroscopic scales; no two experiments are truly identical and what one gets is a function of the experiment. The applicability of our constructions is plausible even if we do not know how to formalize the underlying probability space.

We now derive some properties of the tensors $\mathcal{R}_{ij}, \Phi_{ij}$, and some related tensors for flows that satisfy the Navier-Stokes equations.

We take for granted, even in the case of three space dimensions where this has not been established, that the solutions of the Navier-Stokes equations exist for all time and are differentiable. (There is a difficulty involved, since a random field is defined here by its finite dimensional distributions, which do not contain enough information to decide questions regarding the differentiability of solutions.) We are thus assuming that for each ω, the random variables $u_i(\mathbf{x},\omega)$ can be viewed as values at points $\mathbf{x}$ of smooth functions $u_i(\mathbf{x})$.

Since div $\mathbf{u} = 0$, we have

$$\langle u_i(\mathbf{x}) \frac{\partial u_j(\mathbf{x}+\mathbf{r})}{\partial r_j}\rangle = 0 = \frac{\partial}{\partial r_j}\ \mathcal{R}_{ij}(\mathbf{r})\ ;$$

---

[1]For a recent textbook about chaos, see, e.g., S. Wiggins, *Introduction to Applied Dynamical Systems and Chaos*, 1991.

(the summation convention is used). Note that the calculation of means and differentiation commute, as one can readily establish. Similarly,

$$\frac{\partial}{\partial r_i} \mathcal{R}_{ij}(\mathbf{r}) = 0 \; ;$$

thus

$$k_j \Phi_{ij}(\mathbf{k}) = k_i \Phi_{ij}(\mathbf{k}) = 0 \; .$$

Let $\boldsymbol{\xi}(\mathbf{x})$ be the vorticity vector, $\boldsymbol{\xi} = \operatorname{curl} \mathbf{u}$; the components of $\boldsymbol{\xi}$ are given by

$$\xi_i = \varepsilon_{ijk} \frac{\partial u_k}{\partial x_j} \; ,$$

where $\varepsilon_{ijk} = \begin{cases} 0 & i,j,k \text{ are not all different} \\ 1 & \text{if } i,j,k \text{ is an even permutation of 1,2,3} \\ -1 & \text{if } i,j,k \text{ is an odd permutation of 1,2,3} \; . \end{cases}$

We have, writing $\mathbf{x}' = \mathbf{x} + \mathbf{r}$,

$$
\begin{aligned}
\langle \xi_i(\mathbf{x}) \xi_j(\mathbf{x}') \rangle &= \varepsilon_{i\ell m} \varepsilon_{jpq} \left\langle \frac{\partial u_m}{\partial x_\ell}(\mathbf{x}) \frac{\partial u_q}{\partial x'_p}(\mathbf{x}') \right\rangle \\
&= -(\delta_{ij}\delta_{\ell p}\delta_{mq} + \delta_{ip}\delta_{\ell q}\delta_{mj} + \delta_{iq}\delta_{\ell j}\delta_{mp} \\
&\quad - \delta_{ij}\delta_{\ell q}\delta_{mp} - \delta_{ip}\delta_{\ell j}\delta_{mq} - \delta_{iq}\delta_{\ell p}\delta_{mj}) \frac{\partial^2}{\partial r_\ell \partial r_p} \mathcal{R}_{mq}(\mathbf{r})
\end{aligned}
$$

$$(2.5) \qquad = -\delta_{ij}\Delta\mathcal{R}_{\ell\ell} + \frac{\partial^2 \mathcal{R}_{\ell\ell}}{\partial r_i \partial r_j} - \Delta\mathcal{R}_{ji}(\mathbf{r}) \; ,$$

where Δ is the Laplace operator $\Sigma\partial_j^2$. Thus,

$$\langle \xi_i(\mathbf{x})\xi_i(\mathbf{x}+\mathbf{r}) \rangle = -\Delta\mathcal{R}_{ii}(\mathbf{r}) \; ,$$

and in particular

$$\langle \xi_i(\mathbf{x})\xi_i(\mathbf{x}) \rangle = \langle \boldsymbol{\xi}(\mathbf{x}) \cdot \boldsymbol{\xi}(\mathbf{x}) \rangle = -\Delta\mathcal{R}_{ii}(\mathbf{r})|_{\mathbf{r}=0} = \int (+k^2)\Phi_{ii}(\mathbf{k})d\mathbf{k} \; .$$

If we integrate the right-hand side over the sphere of radius $|\mathbf{k}| = k$, we obtain

$$\langle \boldsymbol{\xi}(\mathbf{x}) \cdot \boldsymbol{\xi}(\mathbf{x}) \rangle = \int_0^\infty Z(k)dk \; ,$$

where

$$(2.6) \qquad Z(k) = \int_{|\mathbf{k}|=k} k^2 \Phi_{ii}(\mathbf{k})d\mathbf{k} \qquad \text{and} \qquad Z(k) = k^2 E(k) \; .$$

$Z(k)$ is the vorticity spectrum, and is thus related in a very simple way to the energy spectrum.

We now want to calculate the dissipation spectrum, i.e., the contribution of various wave numbers to the total energy dissipation.

The Navier-Stokes equations read, in component form,

$$\partial_t u_i + u_k \partial_k u_i = -\partial_i p + R^{-1} \Delta u_i \ ,$$

where $u_i = u_i(\mathbf{x})$, $\partial_i \equiv \frac{\partial}{\partial x_i}$, and the summation convention is used. Let $\mathbf{x}' = \mathbf{x} + \mathbf{r}$, and write $u_i' = u_i(\mathbf{x}')$ for brevity. The equations at $\mathbf{x}'$ read

$$\partial_t u_j' + u_k' \partial_k u_j' = -\partial_j p' + R^{-1} \Delta_{x'} u_j' \ ;$$

the forces $\mathbf{f}$ have been omitted because they play no role in the argument. Assume also that $\langle \mathbf{u} \rangle = 0$.

Multiply the equation at $\mathbf{x}$ by u_j' and the equation at $\mathbf{x}'$ by u_i; the result is

(2.7) $$u_j' \partial_t u_i + u_j' u_k \partial_k u_i = -u_j' \partial_i p + R^{-1} u_j' \Delta u_i \ ;$$

Similarly,

(2.8) $$u_i \partial_t u_j' + u_i u_k' \partial_k u_j' = -u_i \partial_j p' + R^{-1} u_i \Delta_{x'} u_j' \ .$$

Now average the equations, i.e., take the expected value of each term and add the results; note that

$$\langle u_j' \partial_t u_i + u_i \partial_t u_j' \rangle = \partial_t \mathcal{R}_{ij} \ .$$

Furthermore,

$$\partial_{x_i'} \equiv \frac{\partial}{\partial x_i'} = \partial_{r_i} = \frac{\partial}{\partial r_i} \ ;$$

$$\partial_{x_i} \equiv \frac{\partial}{\partial x_i} = -\partial_{r_i} = -\frac{\partial}{\partial r_i} \ .$$

Note that for the average of homogeneous random function integration by parts is legitimate, e.g.,

$$\partial_x(uv) = u \partial_x u + v \partial_x u \ ,$$

taking averages, and noting that by homogenity,

$$\langle \partial_x(uv) \rangle = \partial_x \langle uv \rangle = 0 \ ,$$

we obtain

$$\langle u \partial_x v \rangle = -\langle v \partial_x u \rangle \ .$$

Setting $i = j$, and applying the appropriate integrations by parts to equations (2.7) and (2.8) we find

$$\partial_t \mathcal{R}_{ii}(\mathbf{r}) = -\partial_{x_k}\langle u_i' u_k u_i' - u_i u_k' u_i\rangle - \partial_{x_i}\langle p u_i' - p' u_i\rangle + \frac{2}{R}\Delta_{\mathbf{r}}\mathcal{R}_{ii}(\mathbf{r}) \ ,$$

where $\Delta_{\mathbf{r}} = \Sigma\partial_{r_j}^2$. All the terms on the right-hand side except for the last one vanish at $\mathbf{r}=0$ by homogeneity, thus

$$\partial_t \mathcal{R}_{ii}|_{\mathbf{r}=0} = \frac{d}{dt}\langle \mathbf{u}^2\rangle = \frac{2}{R}\Delta_{\mathbf{r}}\mathcal{R}_{ii}|_{\mathbf{r}=0} = -\frac{2}{R}\int k^2\Phi_{ii}(\mathbf{k})d\mathbf{k} \ .$$

$2R^{-1}k^2\Phi_{ii}$ is the contribution of the motion with wave number around $\mathbf{k}$ to the total dissipation. Integrating this expression on the sphere of radius $k = |\mathbf{k}|$ first, we obtain

$$\partial_t \mathcal{R}_{ii}|_{\mathbf{r}=0} = \frac{d}{dt}\langle \mathbf{u}^2\rangle = -\int_0^\infty D(k)dk \ ,$$

where $D(k) = 2R^{-1}k^2 E(k)$ is the dissipation spectrum. The vorticity and dissipation spectra are proportional to each other. It does not follow that dissipation and vorticity Are themselves proportional. Two very different functions, with different supports, can have the same spectrum. We have noted before that the spectrum provides only partial information about a random function.

2.4. Random Fourier Transform of a Homogeneous Flow Field

We have, in the last few sections, used a Fourier transform of correlation functions and used them to define energy, vorticity and dissipation spectra. Can one also find a Fourier transform of the homogeneous flow field itself? Such a Fourier transform cannot be a usual Fourier transform, because a sample of a homogeneous flow field is not likely to decay at infinity so that a usual Fourier transform would make sense. However, a random Fourier transform can be defined in a sensible manner. More generally, we shall be considering representations of the form

$$\mathbf{u}(\mathbf{x}, \omega) = \int g(\mathbf{x}, \mathbf{s})\rho(d\mathbf{s}),$$

where $g(\mathbf{x}, \mathbf{s})$ is a non-random kernel and $\rho(d\mathbf{s})$ is a random quantity with simple properties. The special case $g(\mathbf{x}, s) = e^{i\mathbf{k}\cdot\mathbf{s}}$ will be the random (or generalized) Fourier transform. The condition for its existence is that $\langle \mathbf{u}^2(\mathbf{x}, \omega)\rangle$ be finite for each $\mathbf{x}$, i.e., that the average energy at a point be finite (as one could expect in fluid mechanics). By contrast, for the usual Fourier transform one must require that the energy in the whole space be finite.

Our discussion will be descriptive; for full detail, see, e.g., Gikhman and Skorokhod.[2]

Let $(\Omega, \mathcal{B}, P)$ be a probability space, and let $L_2(\Omega, \mathcal{B}, P)$ be the space of random variables η defined on Ω that have finite first and second moments, $|\langle \eta \rangle|, \langle \eta^2 \rangle < +\infty$. The inner product of two such variables is defined as $\langle \eta_1 \eta_2 \rangle$, and the norm of η is $\sqrt{\langle \eta^2 \rangle}$.

Let $\mathcal{D}$ be a region in a finite dimensional space; we are going to establish a one-to-one relationship between random variables in $L_2(\Omega, \mathcal{B}, P)$ and functions that are square integrable in $\mathcal{D}$ in a certain sense. To do this, we construct $\mathcal{A}$, an algebra of subsets of $\mathcal{D}$ (an algebra is a collection of sets such that if $A_1, A_2 \in \mathcal{A}$, then CA_1, the complement of $A_1, A_1 \cap A_2$, $A_1 \cup A_2$, all belong to $\mathcal{A}$). A σ-algebra is of course an algebra; note that we have two algebras of sets in play, one on Ω and one on $\mathcal{D}$.

To every A in $\mathcal{A}$ we assign a random variable $\rho = \rho(A, \omega)$, $\omega \in \Omega$, such that

(i) $\rho(A) \in L_2(\Omega, \mathcal{B}, P)$, $\rho(\emptyset) = 0$;
(ii) $\rho(A_1 \cup A_2) = \rho(A_1) + \rho(A_2)$ if $A_1 \cap A_2 = \emptyset$,
(iii) $\langle \rho(A_1) \rho^*(A_2) \rangle = m(A_1 \cap A_2)$,

m is a set function by construction (i.e., a rule that assigns a number to each A in $\mathcal{A}$). We shall call the family of random variables $\rho(A)$ an elementary random measure (the word "elementary" here is a reminder that not all of the axioms that define a measure have yet their appropriate random analogues), and we shall call $m(A)$ the structure function of $\rho(A, \omega)$.

$m(A)$ is non-negative by (iii):

$$m(A) = m(A \cap A) = \langle \rho(A) \rho^*(A) \rangle = \langle |\rho(A)|^2 \rangle > 0 \; ;$$

We also have $m(\emptyset) = 0$ by (i), and if $A_1 \cap A_2 = \emptyset$,

$$
\begin{aligned}
m(A_1 \cup A_2) &= \langle |\rho(A_1) + \rho(A_2)|^2 \rangle \\
&= m(A_1) + m(A_2) + 2m(A_1 \cap A_2) = m(A_1) + m(A_2) \; .
\end{aligned}
$$

This property is called "finite additivity". One can show (and we shall not do so) that a family $\rho(A)$, $A \subset \mathcal{A}$ can be constructed.

Let A be a set in $\mathcal{A}$; its characteristic function is

$$\chi_A(\mathbf{x}) = \begin{cases} 1 & \mathbf{x} \in A \\ 0 & \mathbf{x} \notin A \; . \end{cases}$$

[2]V. Gikhman and A. Skorokhod, *Introduction to the Theory of Random Processes*, 1969; also A. Chorin, *Lectures on Turbulence Theory*, Publish or Perish, 1975.

Let $A_1, \ldots, A_n$ be a finite family of disjoint sets in $\mathcal{A}$, whose union is the whole space $\mathcal{D}$, and construct the function

$$q(\mathbf{x}) = \sum_{k=1}^{n} c_k \chi_{A_k}(\mathbf{x}) ; \qquad c_k = \text{constants} .$$

Such a function $q(\mathbf{x})$ is called "simple". $q(\mathbf{x})$ is defined for all $\mathbf{x}$. To each such $q(\mathbf{x})$ one associates a random variable

$$(2.9) \qquad \eta = \sum_{k=1}^{n} c_k \rho(A_k, \omega) ,$$

where $\rho(A_k)$ is the random variable associated with A_k.

Let q_1, q_2 be two simple functions,

$$q_1(\mathbf{x}) = \sum_{1}^{n} c_k \chi_{A_k}(\mathbf{x}) ,$$

$$q_2(\mathbf{x}) = \sum_{1}^{n} d_k \chi_{A_k}(\mathbf{x}) ;$$

by judicious use of intersections one can use the same A_k's in both definitions, as we have done. Define the inner product of q_1, q_2 by

$$(q_1, q_2) = \sum c_k d_k^* m(A_k) ;$$

the right-hand side can be considered as an integral

$$\int q_1(\mathbf{x}) q_2^*(\mathbf{x}) m(d\mathbf{x}) .$$

Note that $(q_1, q_2) = \langle \eta_1 \eta_2 \rangle$, where η_1, η_2 are the random variables associated with q_1, q_2:

$$\langle \eta_1 \eta_2 \rangle = \left\langle \left(\sum c_k \rho(A_k) \right) \left(\sum_{k} d_k \rho(A_k) \right)^* \right\rangle = \sum c_k d_k m(A_k) .$$

The mapping $q(\mathbf{x}) = \sum a_k \chi_{A_k} \to \sum a_k \rho(A_k)$ can be extended to all square integrable functions with respect to a measure generated by $m(A_k)$ [i.e., obtained by extending $m(A_k)$ to a σ-algebra] and to all random variables η with $\langle \eta^2 \rangle < +\infty$ on $(\Omega, \mathcal{B}, P)$, and it is then one-to-one and onto. Thus, to each function q defined on $\mathcal{D}$ that is square integrable with respect to the appropriate measure one can associate a random variable η with finite variance, and vice versa; symbolically,

$$(2.10) \qquad \eta(\omega) = \int q(\mathbf{x}) \rho(d\mathbf{x}) ,$$

with

$$\langle \eta^2 \rangle = \int q^2(\mathbf{x}) m(d\mathbf{x}) , \qquad m(d\mathbf{x}) = \langle |\rho(d\mathbf{x})|^2 \rangle .$$

A random field $\mathbf{u}(\mathbf{x}, \omega)$ is a collection of random variables, one per value of $\mathbf{x}$. The natural extension of (2.10) to $u(\mathbf{x}, \omega)$ would be

(2.11)
$$u(\mathbf{x}, \omega) = \int g(\mathbf{x}, \mathbf{s}) \rho(d\mathbf{s}) ,$$

where $g(\mathbf{x}, \mathbf{s})$ is a function of two variables that has the same role as q in (2.10); equality here must be interpreted in the L_2 sense, i.e., $u = \int g(\mathbf{x}, \mathbf{s}) \rho(d\mathbf{s})$ means

$$\left\langle \left| u - \int g\rho(d\mathbf{s}) \right|^2 \right\rangle = 0 .$$

Suppose $u(\mathbf{x}, \omega)$ can be represented as in equation (2.11); assume for simplicity that $\langle u \rangle = 0$. Then the correlation function $\mathcal{R}(\mathbf{x}_1, \mathbf{x}_2)$ is

$$
\begin{aligned}
\mathcal{R}(\mathbf{x}_1, \mathbf{x}_2) &= \langle u(\mathbf{x}_1, \omega) u^*(\mathbf{x}_2, \omega) \rangle \\
&= \left\langle \int g(\mathbf{x}_1, \mathbf{s}) \rho(d\mathbf{s}) \int g(\mathbf{x}_2, \mathbf{s}) \rho(d\mathbf{s}) \right\rangle \\
(2.12) \qquad &= \int g(\mathbf{x}_1, \mathbf{s}) g^*(\mathbf{x}_2, \mathbf{s}) m(d\mathbf{s}) ,
\end{aligned}
$$

where $m(d\mathbf{s})$ is the appropriate measure on $\mathcal{D}$. It turns out that the converse is true: If the correlation function $\mathcal{R}(\mathbf{x}_1, \mathbf{x}_2)$ of a random field $u(\mathbf{x}, \omega)$ has the form

$$\mathcal{R}(\mathbf{x}_1, \mathbf{x}_2) = \int g(\mathbf{x}_1, \mathbf{s}) g^*(\mathbf{x}_2, \mathbf{s}) m(d\mathbf{s})$$

for some measure $m(d\mathbf{s})$, then there exists a stochastic measure $\rho(A)$, $A \in \mathcal{A}$, a σ-algebra of sets in $\mathcal{D}$, such that

$$u(\mathbf{x}) = \int g(x, s) \rho(d\mathbf{s}) ;$$

$\langle |\rho^2(d\mathbf{s})|\rangle = m(d\mathbf{s})$, and this equality holds with probability 1 for every $\mathbf{x}$ in $\mathcal{D}$. We shall not prove this converse here, but shall use it anyway.

As a first application, consider a homogeneous flow field. For the sake of economy in notation, u and x will be written as scalar. We know

$$\mathcal{R}(x_1, x_2) = \int_{-\infty}^{+\infty} e^{ik(x_1 - x_2)} dF(k) .$$

The function $F(k)$ defines a measure on the line. Write $g(x,s) = e^{isx}$, $m(ds) = dF(s)$; then

$$u(x,\omega) = \int e^{ikx} \rho(dk) , \qquad \langle |\rho(dk)|^2 \rangle = dF(k) .$$

This is the random Fourier transform, which exists whenever $\langle |u^2(x,\omega)| \rangle$ is finite, and generalizes the usual Fourier transform. The energy spectrum is given by

$$E(k)dk = \left\langle \int_{k<|\mathbf{k}|<k+dk} |\rho^2(\mathbf{k})| m(d\mathbf{k}) \right\rangle ,$$

a straightforward generalization of equation (1.32); it verifies in particular that $E(k) \geq 0$. If u is a vector $\mathbf{u}$, ρ is a vector $\boldsymbol{\rho}$, and the equations have to be suitably reinterpreted.

Another useful representation for $u(x,\omega)$ can be obtained as follows: assume $F'(k) = \phi(k)$ exists, $dF(k) = \phi(k)dk$; $\phi(k) \geq 0$. Let $\hat{h}(k) = \sqrt{\phi(k)}$ be a square root of ϕ; such a square root is not uniquely defined, since one can take the positive or the negative square root at different values of k. Then

$$\mathcal{R}(x_1, x_2) = \mathcal{R}(x_1 - x_2) = \int_{-\infty}^{+\infty} e^{ikx_1} \hat{h}(k) e^{-ikx_2} \hat{h}(k) dk .$$

Remember that if $q_1(x), q_2(x)$ are functions such that

$$\int q_1^2(x)dx < +\infty , \qquad \int q_2^2(x)dx < +\infty ,$$

and $\hat{q}_1, \hat{q}_2$ are their Fourier transforms, then

$$\int q_1(x)q_2(x)dx = \int \hat{q}_1(k)\hat{q}_2^*(k)dk$$

(the Fourier transform preserves inner products); furthermore, the Fourier transform of $q_1(x + a)$ is

$$\frac{1}{\sqrt{2\pi}} \int e^{ikx} q_1(x+a)dx = \frac{1}{\sqrt{2\pi}} \int e^{ik(x-a)} q_1(x)dx = e^{-ika} \hat{q}_1(k) .$$

Thus

$$\mathcal{R}(x_1, x_2) = \int h(x - x_1)h(x - x_2)dx ,$$

where h is the inverse Fourier transform of $\hat{h}$. An application of the theorem above, with

$$\begin{aligned} g(x,s) &= h(s - x) , \\ m(ds) &= ds , \end{aligned}$$

yields

$$u(x) = \int h(x-s)\rho(ds) \ ,$$

where $\langle |\rho(ds)|^2 \rangle = ds$ and, as always, $\langle \rho(A_1)\rho(A_2) \rangle = 0$ if $A_1 \cap A_2 = \emptyset$.

This representation is called the moving average representation of a random field. Indeed, the random integral can be approximated by

$$\sum_{-\infty}^{+\infty} h(x-s_i)\rho(ds_i) \ ,$$

where the $\rho(ds_i)$ are orthogonal random variables (which, of course, are not necessarily independent). This looks like a sum of translates of a single function, with random coefficients. One should be careful to note that neither h nor, as a consequence, $\rho(ds)$, is unique, and that interpretations of this representation require care. We shall find this formula useful in the discussion of two-dimensional vortex motion.

As far as the random Fourier representation is concerned, it is good to know that it exists, and then that it makes sense to speak of the Fourier representation of a homogeneous flow field. In practice, the expressions one obtains for the random Fourier transform $\rho(dk)$ look very much like what one obtains for the coefficients of an ordinary Fourier series in a periodic domain, and one typically manipulates the latter while keeping the former in mind.

2.5. Brownian Motion and Brownian Walks

We now present a brief introduction to Brownian motion and Brownian walks. This section fits here because it is a logical extension of the discussion of random fields in Section 2.2 and because the results will be needed soon. It has no logical connection with the preceding two sections.

A random field is defined as a random function of position: $u(x,\omega)$, $\omega \in \Omega$, x belonging to some region in physical space. Brownian motion is usually presented as an example of a random function of time: $w(t,\omega)$, where $\omega \in \Omega$, Ω a probability space, and t is time. For historical reasons there is a change in vocabulary when one shifts to functions of time: "random field" becomes "stochastic process", "homogeneous" becomes "stationary". Thus a stationary stochastic process is a random function of time whose statistics are invariant under a shift in time.

Brownian motion is a (non-stationary) stochastic process $w(t,\omega)$ such that:

(i) $w(0,\omega) = 0$ for all ω (i.e., all Brownian paths start at the origin; a Brownian path is a sample of Brownian motion).

(ii) If $0 < t_1 < t_2 < t_3 < t_4$, then the random variables $w(t_2) - w(t_1)$ and $w(t_4) - w(t_3)$ are independent.

(iii) For all $s, t \geq 0$, the variable $w(t + s) - w(t)$ is a gaussian variable with mean 0 and variance $s/2$.

(iv) With probability one, $w(t, \omega)$ is a continuous function of t for each ω.

One has to show that a process that satisfies these conditions exists, which is not obvious. For example, if (iv) were replaced by the condition that $w(t, \omega)$ be differentiable, then there would be no way to construct the corresponding process. For a proof of existence, see, e.g., Lamperti.[3] A summary of conditions (i)–(iv) is that Brownian motion is a continuous process with independent gaussian increments.

With probability 1, a Brownian path is not a differentiable function of time; heuristically, consider the variable $w(t+s) - w(t)$; its variance is $s/2$; its standard deviation, which is a measure of its order of magnitude, is $\sim \sqrt{s}$, and thus the derivative of w at t behaves like the limit of $\sqrt{s}/s = s^{-1/2}$ as $s \to 0$.

One can derive an interpolation formula for Brownian paths: suppose we consider Brownian motion and measure $w(t_1), w(t_2)$, $t_2 - t_1 > 0$. What can we say about $w(s)$, $t_1 < s < t_2$? The interpolation formula[4] says that

$$w(s) = w(t_1) + (s - t_1)[w(t_2) - w(t_1)]/(t_2 - t_1) + [(t_2 - s)(s - t_1)/(t_2 - t_1)]^{\frac{1}{2}} W,$$

where W is a gaussian variable with mean 0 and variance $\frac{1}{2}$. One can readily see that for $s = t_1$, $w(s) = w(t_1)$, and for $s = t_2$, $w(s) = w(t_2)$; the first two terms just interpolate linearly between $w(t_1)$ and $w(t_2)$; the final term, that measures the departure of $w(s)$ from a linear interpolant, is large compared to $t_2 - t_1$ for $t_2 - t_1$ small. Thus Brownian motion can do wild things in the small, as one could indeed deduce from its non-differentiability.

A special case of condition (iii) above is the fact that $w(t)$ is a gaussian variable with mean 0 and variance $t/2$, i.e., with probability density function $(\pi t)^{-(1/2)} \exp(-x^2/t)$. This is the Green's function for the heat equation $\partial_t v = \frac{1}{2} \partial_x^2 v$, i.e., its solution with the initial data $v(x, 0) = \delta(x)$, $\delta =$ Dirac delta. In other words, if one allows particles to leave the origin in an (x, t) plane and wander about by Brownian motion, $x(t) = w(t)$, the density of these particles will trace out the Green's function of the heat equation as t unfolds. As a consequence, if one wants to solve the heat equation $\partial_t v = \frac{1}{2} \partial_x^2 v$ with data $v(x, 0) = g(x)$, where $g(x) \geq 0$, $\int g(x) dx = 1$ (these conditions can be achieved by a linear change of variables), all one has to do is sprinkle particles on the x axis with density $g(x)$, and then allow

---

[3] J. Lamperti, *Probability*, Benjamin, New York, 1966.
[4] P. Levy, *Le Mouvement Brownian*, Gauthiers-Villars, Paris, 1954.

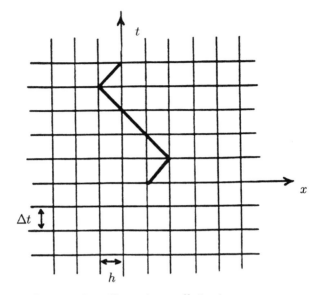

FIGURE 2.1. Brownian walk in time.

these particles to perform independent Brownian motions. The consequent density of the particles at time t will be $v(x,t)$.

Now, give the particles, in addition to the random motion, a steady velocity a (a "drift"); symbolically,

$$(2.13) \qquad dx = adt + dw \ ;$$

in a step of length Δt, Δt small, a particle will move from x to $x + a\Delta t + W$, where W is a gaussian variable with mean 0 and variance $\Delta t/2$. Equation (2.13) was written in a peculiar form in terms of differentials, since, as we have seen, dw/dt does not exist as a regular function. (It does exist as a distribution, and is called "white noise".) The equation satisfied by the density of the particle is now $\partial_t v = a\partial_x v + \frac{1}{2}\partial_x^2 v$.

More generally, if one considers the inviscid vorticity equation in two space dimensions:

$$\partial_t \xi + (\mathbf{u} \cdot \boldsymbol{\nabla})\xi = 0 \ , \qquad \xi(\mathbf{x}, 0) \text{ given} \ ,$$

and one approximates it by a discrete vortex approximation:

$$\xi = \Sigma \xi_i \ , \qquad \mathbf{x}_i \text{ center of supp } \xi_i \ ,$$

$$\frac{d\mathbf{x}_i}{dt} = \mathbf{u}(\mathbf{x}_i) = (K * \xi)(\mathbf{x}_i) \ ,$$

then one can obtain the solution of the Navier-Stokes equation simply by replacing the last equation with

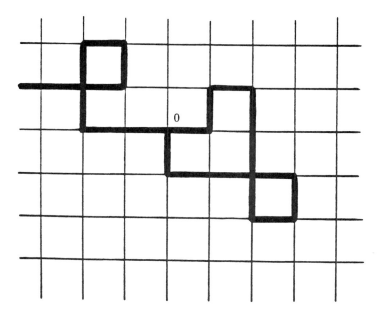

FIGURE 2.2. Brownian walk in space.

$$(2.14) \qquad d\mathbf{x}_i = \mathbf{u}(\mathbf{x}_i)dt + \sqrt{\frac{2}{R}}\; d\mathbf{w}\;,$$

where $\mathbf{w}$ is a two-dimensional brownian motion: $\mathbf{w}(t) = (w_1(t), w_2(t))$, w_1, w_2 two ordinary Brownian motions, independent of each other. The coefficient $\sqrt{\frac{2}{R}}$ is here to adjust the amplitude of $\mathbf{w}$ to what is needed to approximate the diffusion part of the Navier-Stokes equation $D\xi/Dt = R^{-1}\Delta\xi$. Since $\mathbf{u}$ depends on ξ, the particle paths are no longer independent even though the Brownian pushes given to each particle are.[5] This construction can be extended to three space dimensions in several ways; the easiest to describe is the replacement of $\frac{d\mathbf{x}_i}{dt} = \mathbf{u}(\mathbf{x}_i)$ for the Buttke loops (1.28) by an equation of the form (2.14) above.

The discretization of equation (2.14) leads to the random vortex approximation, of which we shall need below only the principle but not the details. The key observation is that the diffusion term in the Navier-Stokes equation can be approximated by simply giving vortex particles random pushes of the right amplitude, the amplitude decreasing with the Reynolds number R.

---

[5] A. Chorin, 1973; J. Goodman, 1987; D. G. Long, 1988.

Consider now a one-dimensional lattice: $x_i = ih$. Suppose a walker starts from 0 and jumps from lattice point to lattice point; each step has length h, to the right with probability $\frac{1}{2}$ and to the left with probability $\frac{1}{2}$, with one step per time interval Δt. We shall call such a walk a Brownian walk. In the (x, t) plane, the walk will look as in Figure 2.1. There are obvious analogues in two space dimensions, with four directions to choose from at each step, and in three space dimensions, with six directions to choose from. As $h \to 0$, these walks converge to a Brownian motion, by the central limit theorem, which asserts that sums of random variables converge to gaussians. The limit is $w(t)$ with a factor that depends on Δt and h.

In Figure 2.2 we display a Brownian walk on a two-dimensional lattice, with the time axis suppressed. In all the applications below the time axis will similarly be suppressed. One can view this suppression as a projection of the walk on the $\mathbf{x}$ plane (or, in three dimensions, on the $\mathbf{x}$ space). One important feature of such walks is that they are not evenly spread in space. After a large but finite number of steps, there are areas with many steps and areas with few steps. There is no impediment to a location being visited twice or, indeed, many times.

3
The Kolmogorov Theory

Chapter 1 summarized some basic results in fluid mechanics, and Chapter 2 presented Fourier analysis tools for describing turbulence. We have not yet dealt with the physics of turbulence. We begin to do this in this chapter, by presenting a version of Kolmogorov's qualitative theory, pointing out both its importance and its vagueness, and laying the groundwork for the developments in later chapters.

3.1. The Goals of Turbulence Theory: Universal Equilibrium

The first statement usually made about turbulent flow is that it involves many scales of motion. What is meant by "scale" is hard to define precisely. One could say that a component of the flow field (an "eddy") has scale L if its Fourier transform has a peak around $k = 1/L$; the problem with this statement is that it is difficult to isolate an appropriate "eddy". One could give a definition of a scale L in terms of what could be seen through a square window of size L^2; this definition also has obvious problems. We shall see later that the problem of defining "scale" is difficult for very precise reasons. However, the intuitive idea is clear: looking at a weather map one sees low or high pressure areas that span continents and storms that cover cities; walking down the street one feels gusts of wind on a human scale. Thus, one observes motions, or "eddies", of various scales.

All these scales of motion are strongly coupled, i.e., to calculate any of them, in particular the larger ones, one has to take into account either all of them or at least a larger fraction of them than may seem reasonable a priori. This can be checked by computation. A calculation on a finite difference grid of mesh size h fails to represent motion on scales comparable with h or smaller. Unless h is extraordinarily small, the resulting approximations to the Navier-Stokes equations for large R do not yield the right answers.

A major goal of turbulence theory is to provide an intuitive explanation of what happens in a turbulent flow. A possibly more important goal, probably related to the first, is to provide some way to calculate accurately without explicitly representing all the interacting scales. The large scales, which are determined by the forces, boundaries, etc., specific to each problem, must presumably be calculated from first principles in each problem. On the other hand, if there are features of the smaller scales that are common to all turbulent flows, as seems to be the case, then an understanding of these small-scale features and a useful representation of them may well be of great help in computation. Thus turbulence theory concentrates on the small scales and seeks to identify and analyze their common features.

The analysis of the small scales of motion in turbulence involves two assumptions. The first is the universal equilibrium assumption: the characteristic time of small-scale motion is small compared with the characteristic time of overall decay. We shall explain this assumption in words before attempting to give it an analytical content. Assume the fluid is subjected to forces which change appreciably over a time T. The time it takes the fluid to respond to the change in the force is comparable to T; thus, the analysis of the response of the large-scale motion to the external forces requires a solution of the equations of motion and does not have a universal character. Suppose there are no forces. The fluid flow will decay to zero. The time it takes the motion to adjust internally is comparable with ℓ/u, where ℓ is a typical length and u is a typical velocity for the problem at hand. This time is found experimentally to be comparable with the time it takes the motion to decay. Thus, the large-scale motion depends strongly on the particular data and geometry of the specific problem being solved, and is not an appropriate subject for an analysis based on problem-independent statistical assumptions. On the other hand, the small-scale motion has a smaller distance to cover in order to adjust itself to changing circumstances; such adjustment can take place in a short time compared to the time it takes the flow to decay, and thus it can conceivably give rise to features independent of the particular problem at hand.

Let us temporarily identify Fourier components of a velocity field with "eddies", i.e., recognizably organized motions in the interior of the fluid. Assume $\mathbf{u}(\mathbf{x})$ is periodic, with Fourier series

$$\mathbf{u}(\mathbf{x}) = \sum \hat{\mathbf{u}}_{\mathbf{k}} e^{i\mathbf{k}\cdot\mathbf{x}};$$

$\hat{\mathbf{u}}_{\mathbf{k}}$ can be associated with eddies of size k^{-1}, $k = |\mathbf{k}|$. The characteristic velocity of the flow is $\langle \mathbf{u}^2 \rangle^{1/2} = u$; the characteristic time of decay of the flow is $u/|du/dt|$; the characteristic time of the small eddies is $1/(ku_k)$, where u_k is an amplitude of $\hat{\mathbf{u}}_{\mathbf{k}}$ for $|\mathbf{k}| = k$ large enough, and the assumption above reads

$$\frac{1}{ku_k} \ll \frac{u}{\left| \frac{du}{dt} \right|} \ .$$

It is of course merely an assumption.

This assumption of universal equilibrium (whose relation to other notions of equilibrium, such as thermal or statistical equilibrium and mechanical equilibrium, will be discussed in later chapters) asserts that the small scales reach an asymptotic state. However, that state could be different from problem to problem. One then needs an assumption about the independence of the small scales and the large scales to assert that the asymptotic state is not problem-dependent and thus possibly common to all problems. We shall see in later chapters that under some conditions this independence of small scales and large scales can be justified but that there are circumstances where it is plainly false. In classical (non-quantum) fluids in three space dimensions the independence condition will turn out to be very plausible, and for the moment we accept it without further analysis. It will also be necessary to explain how the independence of small scales and large scales can be compatible with the strong coupling between scales.

3.2. Kolmogorov Theory: Dimensional Considerations

Consider a homogeneous flow with energy spectrum $E(k)$ and dissipation spectrum $2R^{-1}k^2E(k)$. The homogeneity assumption is not very severe for small-scale motion since it is plausible that any flow can be viewed as homogeneous on a small enough scale. It is customary in the discussion of Kolmogorov's spectrum to require also that the flow be isotropic, i.e., have statistics invariant under rotation. We are content here to assume that the flow is sufficiently different from two-dimensional flow so as not to have the latter's constants of motion.

When R is very large one must wait for a very large k before the dissipation spectrum is significant. Thus the range of wave number k for which $k^2E(k)$ is significant may be quite large because of the lack of damping. The graphs of $E(k)$ and $k^2E(k)$ may look as in Figure 3.1, and have nearly disjoint supports. k_1 is where $E(k)$ is large; $k_1 \sim 1/L$, where L is typically a scale on which the fluid is stirred. k_2 is where the dissipation is large, and we assume $k_1 \ll k_2$.

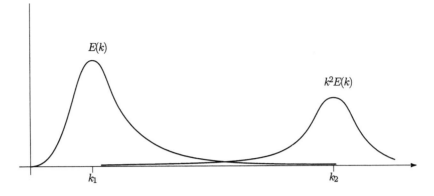

FIGURE 3.1. $E(k)$, $k^2 E(k)$ for large R.

The range between k_1 and k_2 is the "inertial range". One could well believe that all that happens in the inertial range is that energy passes through it from the region around k_1 where the energy is ("the energy range") to the region around k_2 where the energy is dissipated ("the dissipation range"). What could $E(k)$ in this intermediate region depend on? It is plausible that it depends only on k and on the rate of energy dissipation $\epsilon = \frac{d}{dt}\langle \mathbf{u}^2 \rangle$. The units of $E(k)$ are L^3/T^2, where L is a unit of length and T a unit of time [remember $\langle \mathbf{u}^2 \rangle = \int E(k)dk$], k has units L^{-1}, and ϵ units L^2/T^3. The only possible combination that is dimensionally correct is

$$(3.1) \qquad\qquad E(k) = C\epsilon^{2/3} k^{-5/3},$$

where C is a dimensionless constant. This is the famous Kolmogorov law. It is independent of the equations of motion; no mechanical explanation for its validity in three dimensions has yet been offered here. It is well-verified by both physical and numerical experiment.

The dependence of E on ϵ can be easily explained: take the averaged equation of motion for the correlation tensor $\mathcal{R}_{ij}$, set $i = j$, take a Fourier transform, and average over a sphere of radius $|\mathbf{k}| = k$. The result is

$$\partial_t E(k) + 2k^2 R^{-1} E(k) = Q(k),$$

where $Q(k)$ originates from the nonlinear terms in the Navier-Stokes equations. The term $Q(k)$ is cubic in $\hat{\mathbf{u}}(\mathbf{k})$, and since it is the transform of a product of terms, it involves values of $\hat{\mathbf{u}}(\mathbf{k}')$ for various $\mathbf{k}' \neq \mathbf{k}$. It is the term responsible for the energy transfer between different wave numbers, since neither of the terms on the left is. If the energy transfer is reasonably local in $\mathbf{k}$-space, i.e., energy moves from wave number to wave number in the inertial range and not directly from the neighborhood of k_1 to the neighborhood of k_2, then $Q \sim \epsilon \sim u^3$, where u is an amplitude $\hat{\mathbf{u}}$

in the inertial range. Since $E \sim u^2$, then $E \sim \epsilon^{2/3}$, as predicted in the Kolmogorov law. The relations in the preceding sentences are not dimensionally correct, since the proportionalities may contain coefficients that are not dimensionless, as can be seen by writing out Q in full.

One can derive a simple estimate of k_2, a wave number characteristic of the dissipation range. k_2 depends of course on the viscosity ν (in doing dimensional analysis, we must for a moment abandon the practice of burying ν in the dimensionless Reynolds number), and on ϵ [since $E(k)$ depends on ϵ]. The only combination of these parameters that has the dimension of length is $(\nu^3/\epsilon)^{1/4} = \eta$; η is the Kolmogorov scale. It is plausible, and indeed verified by experiment, that η is the approximate order of magnitude of the scales on which dissipation becomes important. Thus $k_2 \sim 1/\eta$, and in particular, $k_2 \to \infty$ as $\nu \to 0$.

The significant assumption in the dimensional analysis that led to the Kolmogorov law (3.1) was that $E(k)$ did not depend on the viscosity ν. As $\nu \to 0$, $E(k)$ in the inertial range keeps its shape, while the dissipation range recedes to the right in Figure 3.1. A reasonable way to understand this assumption is to consider $E_\nu(k)$, the energy spectrum for a flow that is a solution of the Navier-Stokes equations with viscosity ν, and assume that $\lim_{\nu \to 0} E_\nu(k) \to E(k)$ in some reasonable sense for k large enough. [Note that it is far from obvious that $\lim_{\nu \to 0} E_\nu(k) = E_0(k)$, where E_0 is the corresponding spectrum for the Euler equations.] If indeed $E_\nu(k) \to E(k)$, and we are interested in the inertial range only, we can assume that the limit $\nu \to 0$ has been taken, k_2 is at infinity, and $E(k)$ has an energy containing range in $0 \leq k \leq Ck_1$, $C =$ some constant > 1, and an inertial range for $k \geq Ck_1$.

Pick a large constant K such that Ck_1/K is small, and perform the change of variables $k' = k/K$. In terms of k', $E(k') \cong \langle \mathbf{u}^2 \rangle \delta(k') + C_1 \epsilon^{2/3}(k')^{-5/3}$, where C_1 is a constant, and $\delta(k')$ is a Dirac delta function whose coefficient is determined by the fact that almost all the energy lives below Ck_1. The Fourier transform $\Phi_{ii}(\mathbf{k}')$ of the correlation tensor can be estimated by undoing the averaging over a sphere that leads from Φ_{ii} to E:

$$\Phi_{ii}(\mathbf{k}') \cong \langle \mathbf{u}^2 \rangle \delta(\mathbf{k}') + \epsilon^{2/3} O\left((k')^{-5/3-2}\right),$$

where $\delta(\mathbf{k}') = \delta(k_1)\delta(k_2)\delta(k_3)$, the first term on the right expresses the fact that the energy lives in a neighborhood of the origins in the $\mathbf{k}' = \mathbf{k}/K$ variables, and the O in $O(k^{-11/3})$ is there because the unaveraged Φ_{ii} may depend on $\mathbf{k}$ and not only on $k = |\mathbf{k}|$. The $O\left(k^{-11/3}\right)$ term must be smoothed near the origin to make the equation meaningful.

Now take the inverse Fourier transform of this equation. This inverse Fourier transform must be interpreted in the sense of distributions; the smoothing near $\mathbf{k}' = 0$ leads to a smoothing near $\mathbf{r}' = \infty$ ($\mathbf{r}' =$ the variable conjugate to $\mathbf{k}'$) and the two smoothings disappear together. The correct exponents in the inverse Fourier transform can be found by dimensional analysis: $\mathbf{k}'$ has dimension L^{-1} if $\mathbf{r}'$ has dimension L, and the equation that defines the Fourier transform must be dimensionally correct. The result is

$$\langle \mathbf{u}(\mathbf{x} + \mathbf{r}') \cdot \mathbf{u}(\mathbf{x}) \rangle = \langle \mathbf{u}^2 \rangle + O\left(\epsilon^{2/3} (r')^{2/3} \right),$$

where the O in $O\left((r')^{2/3}\right)$ allows for a possible dependence on $\mathbf{r}'$ as well as $|\mathbf{r}'| = r'$. Using $\langle (\mathbf{u}(\mathbf{x} + \mathbf{r}'))^2 \rangle = \langle (\mathbf{u}(\mathbf{x}))^2 \rangle$ (homogeneity) and absorbing a factor of 2 into the O, we find

$$\langle (\mathbf{u}(\mathbf{x} + \mathbf{r}') - \mathbf{u}(\mathbf{x}))^2 \rangle = O\left(\epsilon^{2/3} (r')^{2/3} \right).$$

Let f be a function of $\mathbf{r}$ and $\hat{f}(\mathbf{k})$ its Fourier transform. In three space dimensions, the inverse transform of $\hat{f}(\mathbf{k}/K)$ is $K^3 f(K\mathbf{r})$, i.e.,

(3.2)
$$\widehat{K^3 f(K\mathbf{r})} = \hat{f}(\mathbf{k}/K)$$

Thus $\mathbf{r}' = \mathbf{r}K$. Absorbing powers of K into the O, we find

$$\langle (\mathbf{u}(\mathbf{x} + \mathbf{r}) - \mathbf{u}(\mathbf{x}))^2 \rangle = O(\epsilon^{2/3} r^{2/3}) \quad \text{for small } r;$$

if the flow is isotropic (statistically rotation invariant) as well as homogeneous, this equation can be written as

(3.3)
$$\langle (\mathbf{u}(\mathbf{x} + \mathbf{r}) - \mathbf{u}(\mathbf{x}))^2 \rangle = C_2 \epsilon^{2/3} r^{2/3}, \quad \text{small } r.$$

where C_2 is a constant. This is the Kolmogorov $2/3$ law. The expression on the left-hand side of equation (3.3) is the "second-order structure function". It measures the amplitude of relative motion for $\mathbf{r}$ dual to $\mathbf{k}$ in the inertial range, i.e., on small scales comparable with k^{-1}, where k is in the inertial range. The appearance of the structure function in equation (3.3) can be understood by observing that small "eddies" are mainly transported around by large "eddies" and thus if "eddies" with large wave number contain any energy at all it must be because they have internal motions in addition to the motion that corresponds to their being transported passively.

A differently presented but essentially equivalent derivation of the fact that one can take an inverse Fourier transform of quantities defined for k in the inertial range and obtain physically meaningful results for $\mathbf{r}$ in the inertial range can be found in Batchelor[1]. From now on, we shall take for granted this possibility of taking Fourier transforms of parts of functions. Equation (3.3) can be derived from scratch by dimensional considerations; all one has to do is assume that the second-order structure function depends for r small on ϵ and r only. This derivation obviates the need for

---

[1]G.K. Batchelor, *The Theory of Homogeneous Turbulence*, Cambridge, 1961.

the elaborate analysis of Fourier transforms in this specific case, but the analysis is needed in later applications.

3.3. The Kolmogorov Spectrum and an Energy Cascade

In the last section we derived the Kolmogorov spectrum for the inertial range on the basis of purely dimensional considerations. This derivation, on the face of it, should apply to many nonlinear equations and indeed to the Navier-Stokes equations in two space dimensions, but it does not. In particular, in two space dimensions we know that in a flow with well-behaved vorticity the integral of ξ^2 over space is a constant of the motion. The observations in the previous chapter about the possibility of carrying out integrations by parts for homogeneous random functions show that $\langle \xi^2 \rangle$ is a constant of the motion for random homogeneous flow. Since $\langle \xi^2 \rangle = \int Z(k)dk = \int k^2 E(k)dk$, we cannot have $E(k) \sim k^{-5/3}$ in two dimensions. Thus there must be some mechanical process, specific to three-dimensional flow, which justifies the assumptions in the dimensional analysis that leads to the Kolmogorov law.

The usual mechanical explanation is that the main process in the inertial range is a "cascade" of energy from large-scale motions to small-scale motions. One could think of it as something like a waterfall: energy is fed in at one end of the range and comes out at the other, the flow across the range being irreversible and all in one direction. We shall present this cascade in its extreme "local" form, as presented by Kraichnan.[2]

Divide wave-number space into energy shells, as in Figure 1.3; for example, let the n-th energy shell be $2^n \leq k \leq 2^{n+1}$. Let u_n be a typical velocity associated with this shell (for example, the square root of an average value of a squared Fourier coefficient in the shell); let $\ell_n \sim k_n^{-1}$ be a typical length in the shell, and let $E_n \sim u_n^2$ be the energy associated with the shell, where "$\sim$" denotes not strict equality but equality of orders of magnitude. The characteristic time for motion in the shell, i.e., for motion that corresponds to wave numbers in the shell, is $\ell_n/u_n = \tau_n$. Suppose that in a time τ_n the motion in the shell gives up its energy to other shells; the rate of energy dissipation is then

$$\epsilon \sim \frac{u_n^2}{\tau_n} \sim u_n^2/(\ell_n/u_n) = u_n^3/\ell_n.$$

If ϵ is constant across the spectrum, i.e., if energy is steadily progressing across it, then $u_n^2 \sim \epsilon^{2/3}\ell_n^{2/3}$. If we then identify ℓ_n with r and u_n^2 with $\langle(\mathbf{u}(\mathbf{x}+\mathbf{r}) - \mathbf{u}(\mathbf{x}))^2\rangle$, we have derived the Kolmogorov 2/3 law (3.3) and,

---

[2]R. H. Kraichnan, 1974.

by Fourier transformation, the 5/3 law (3.1). This cascade is "local" because energy moves from each shell to a neighboring shell. We shall see further below that such a cascade is impossible in two dimensions; thus it seems that all is well.

However, it is not at all obvious that this "local" cascade is possible in three space dimensions either; the theory comes out correctly because the only variables in play are u_n, ℓ_n, and ϵ, and thus a dimensionally correct result necessarily leads to the Kolmogorov conclusion. Let us try to sharpen the physical picture implied in the local cascade argument.

It cannot be the case that all the energy moves from one shell to the next shell in a characteristic time. There are two reasons:

(i) If all the energy moved from one shell to the next, the energy would be the same in each shell. Thus $E_n \sim u_n^2 \cong$ constant, contrary to what has been deduced. If u_n^2 decreases with n, were does the extra energy go?

(ii) If the energy moves from one shell to the next, and the energy spectrum has no gaps, the time of transmission τ_n must be the same for all shells, again contrary to what has been deduced.

One could rescue the local cascade picture with energy moving from shell to neighboring shell by making it slightly more complex. One can divide physical space with boxes and take a standard (non-random) Fourier transform in each. One can imagine a local cascade occurring in each box separately, with energy moving from shell to shell. In each shell for a given box $E_n = E$ is a constant; the average energy per shell for the whole flow is thus proportional to the fraction of time the energy spends in each shell:

$$\langle u_n^2 \rangle \sim E \cdot \frac{\tau_n}{T},$$

where $\tau_n = \ell_n/\sqrt{E}$ and T is the time it take the energy to flow across the inertial range. Thus $\langle u_n^2 \rangle \sim \frac{\sqrt{E}}{T}\ell_n = (\epsilon/\sqrt{E})\ell_n$, and by a Fourier transform $E(k) \sim (\epsilon/\sqrt{E}) \cdot k^{-2}$. We do not claim that this is the truth (the Kolmogorov law is well established experimentally) but only demonstrate that a reasonable local cascade picture that violates the assumptions of the Kolmogorov picture can exist and lead to a different spectrum.

If a cascade picture makes sense, one probably must have a complex interplay between distant shells. On the other hand, nothing in the Kolmogorov theory forbids one to consider the establishment of the spectrum as being irreversible, like the spreading of water that has initially been contained in a portion of a tub, but with the flow once the spectrum has been established being nearer to a sloshing to and fro, with some leakage occurring on one end of the tub to account for the over-all energy loss in turbulent flow. In other words, it may well be that once the inertial range has been formed, energy goes back and forth across the spectrum, with the energy dissipation being just the difference between almost equally large energy flows in both directions in wave number space.[3] We shall make an argument in favor of this "bathtub" picture in later chapters. The possibility that the Kolmogorov spectrum is correct but that the arguments that support it are not will have a counterpart in Chapter 6 in the analysis of Flory exponents, where the situation is much clearer.

As has been pointed out recently,[4] the Kolmogorov theory itself contains an argument in favor of the "bathtub" picture. We start with a few generalities. Consider $\mathbf{u}^I(\mathbf{x}, \omega)$, the flow in the inertial range. It can presumably be isolated from the energy and dissipation range by taking a Fourier transform, setting to zero the amplitudes $\rho(d\mathbf{k})$ for $\mathbf{k}$ outside the range, and then taking the inverse Fourier transform. If the inertial range is broad enough in wave number space, $\mathbf{u}^I$ presumably (note the frequency with which this adverb is used) obeys the Euler equations on its own; the viscous terms should be negligible (see the assumptions under which the 5/3 law was derived), and the energy range should be independent of the inertial range (see Section 3.1). The Euler equations are invariant under the time reversal $t \to -t$, $\mathbf{u} \to -\mathbf{u}$. The formation of the inertial range is irreversible, in the following sense: if one starts with a very smooth flow without an inertial range, an inertial range will form. If one starts from a "mature" flow, with an inertial range, one does not expect the inertial range to disappear if the flow evolves by the Euler equations, which hold whether one goes backward or forward in time.

On the other hand, if the cascade picture holds, time reversal will undo the inertial range, because it will make energy go up to the large scales. Thus the invariance of the inertial range under time reversal, suggested by the Euler equations, is incompatible with the cascade picture, unless the existence of the dissipation range dominates the whole spectrum. Thus, once $\mathbf{u}^I$ has developed, if time reversal changes the inertial range the cascade picture is plausible, and if time reversal does not change the inertial range, then the cascade picture is implausible.

The even moments of $\mathbf{u}^I$ are invariant under the reversal $\mathbf{u} \to -\mathbf{u}$ but the odd ones change sign. Thus invariance under time reversal presumably means that the odd moments of $\mathbf{u}^I$ are zero (we shall have much more to say about this conclusion later). If the odd moments of $\mathbf{u}^I$ are indeed zero the "bathtub" picture becomes more plausible.

Consider the third-order structure functions $\langle (u_i(\mathbf{x} + \mathbf{r}) - u_i(\mathbf{x}))^3 \rangle$. If, as is appropriate in the inertial range, they depend only on ϵ and on r, we

---

[3] See also C. Meneveau, 1991.

[4] A manuscript by P. Kailasnath, A. Migdal, K. Sreenivasan, V. Yakhot, and L. Zabair, 1992, as understood by the author.

must have

$$\langle (u_i(\mathbf{x} + \mathbf{r}) - u_i(\mathbf{x}))^3 \rangle = C_{3i}\epsilon r; \quad i = 1, 2, 3,$$

where C_{3i} are constants and r becomes a shorthand for $O(r)$ if the flow is not isotropic. A Fourier transformation then gives

$$\widehat{f_i^3} \equiv \langle \widehat{(\Delta u_i)^3} \rangle = C_{3i}\epsilon \delta'(\mathbf{k}),$$

where $\Delta u_i = u_i(\mathbf{x} + \mathbf{r}) - u_i(\mathbf{x})$ and $\delta'(\mathbf{k})$ is the (distribution) derivative of the delta function $\delta(\mathbf{k})$. Thus $\widehat{f_i^3}$ vanishes everywhere except at the origin, and the third moments of u_i^I are approximately zero through all of the inertial range. This analysis supports the bathtub picture.

One could try to analyze the higher odd moment of $\mathbf{u}^I$; there is however little experimental support for the notion that Kolmogorov scaling describes these higher moments well, so the analysis is unlikely to be meaningful.

This may be the right point to emphasize that some third moments of $\mathbf{u}$ and its derivatives (i.e., expressions such as $\langle u_i u_j \partial_j u_k \rangle$) must be non-zero or else energy transfer between $\mathbf{k}$'s will not occur (see the expression for Q above). We have not made an argument that all third moments vanish, only those that have a bearing on the question of reversibility of the inertial range once it has been established.

3.4. Fractal Dimension

Before proceeding with the discussion of the Kolmogorov spectrum, we need some facts about fractal sets.[5]

Consider a set C contained in an n-dimensional space. Cover it with a finite number of balls of radii ρ_i, $\rho_i \leq \rho$ (assuming that such a cover can be found). Form the sum $S_d(C) = \sum \rho_i^d$, and consider the quantity

$$h_d(C) = \lim_{\rho \to 0} \liminf S_d(C),$$

where the lim inf is taken over all finite covers of C with $\rho_i \leq \rho$. $h_d(C)$ is the Hausdorff measure of C in dimension d.

If C is a reasonable set, for example, a closed bounded set, then $h_d(C)$ is zero for d large enough and infinite for d small enough. For example, if C is the unit ball in two space dimensions, $h_2(C) = \pi^{-1}$; for $d > 2$, $\sum \rho_i^d \leq \rho^{d-2} \sum \rho_i^2$ and it follows that $h_d(C) = 0$; if $d < 2$, take a finite cover, take $\rho_{\min}$, the smallest of the ρ_i, and note that $\sum \rho_i^d \geq \frac{1}{\rho_{\min}^{d-2}} \sum \rho_i^2$, thus $h_d(C) = \infty$. It can be shown that the number D: $D =$ greatest lower bound of d for which $h_d(C) = \infty$, $=$ least upper bound of d for which $h_d(C) = 0$, exists

---

[5] B. Mandelbrot, *Fractals: Form, Chance and Dimension*, Freeman, 1977.

for reasonable C; it is called the Hausdorff dimension of C. The Hausdorff measure $h_D(C)$ of C in dimension D is simply called the Hausdorff measure of C. It may be zero, finite and non-zero, or infinite. Clearly, if C_1, C_2 are reasonable disjoint sets, $h_D(C_1 \cup C_2) = h_D(C_1) + h_D(C_2)$; if all the linear dimensions of the space in which C is imbedded are stretched by a factor ℓ, then $h_D(C)$ is multiplied by ℓ^D: $h_D(\ell C) = \ell^D h_D(C)$. The dimension D of a circle in the plane is 2, of a ball in three dimensions is 3, i.e., for usual objects the Hausdorff dimension coincides with the usual dimension.

As a non-trivial example, consider the Cantor set C: divide $[0,1]$ into three segments of length $1/3$ and throw away the middle one; take each of the remaining pieces and throw away the middle third, and so on. The remainder is the Cantor set. Assume that the Hausdorff measure of C is finite and non-zero. C consists of two subsets of equal measure, each subset is a copy of the whole reduced by a factor $1/3$, thus

$$h_D(C) = 2 \cdot \frac{1}{3^D} \cdot h_D(C),$$

and if $h_D(C)$ is finite, one can deduce $D = \log 2/\log 3$.

Consider further a set C and suppose that its mass m (defined in some appropriate way) satisfies the scaling relation $m(\ell C) = \ell^D m(C)$ when the object is scaled up by a factor of ℓ while the density remains constant. We shall say that C has fractal dimension D. Clearly an object of Hausdorff dimension D has fractal dimension D. A fractal set is a set whose fractal dimension is in some way unusual, for example, non-integer.

As an example, consider a Brownian walk on a two-dimensional square lattice, with the time axis removed (Section 2.5). Assume the lattice bonds have length h. A random walk is the union of the lattice bonds across which the random walker has jumped. Assume there are N steps in a walk. Such a walk can be viewed as an N-dimensional random variable, each one of whose components takes on the values $\pm h$; the components are independent of each other. Consider the end-to-end straight line length of the walk r_N, $r_N^2 = x_N^2 + y_N^2$, where

$$x_N = \sum_{i=1}^{N} \eta_i,$$

$\eta_i = h$ with probability $1/4$, $\eta_i = -h$ with probability $1/4$, $\eta_i = 0$ with probability $1/2$, $\langle \eta_i \eta_j \rangle = 0$ if $i \neq j$, $\mathrm{Var}(\eta_i) = \frac{1}{2}h^2$, $\langle \eta_i \rangle = 0$. By the central limit theorem, the probability density of x_N is a gaussian with mean 0 and variance $\frac{1}{2}Nh^2$, thus

$$\langle x_N^2 \rangle^{1/2} = \tfrac{1}{2}hN^{1/2}, \quad N \text{ large}.$$

Similarly, $\langle y_N^2 \rangle^{1/2} = \frac{1}{2} h N^{1/2}$. x_N and y_N are uncorrelated, and for large N become independent. Hence

$$\bar{r}_N \equiv \langle r_N^2 \rangle^{1/2} = h N^{1/2}$$

or

(3.4) $$N \sim (\bar{r}_N)^2.$$

$\bar{r}_N$ is a measure of the linear extent of the random walk, and thus for large N the random walk behaves like an object of fractal dimension 2. A further calculation that uses the Tchebysheff inequality of Section 2.1 shows that the probability of a long path having a length that differs significantly from $h N^{1/2}$ is small. An appropriate limiting procedure, in which $h \to 0$ while N increases and $\bar{r}_N$ remains fixed, allows us to conclude that, with probability one, the fractal dimension of a Brownian motion path in two-dimensional space is two. Remember that a path is here the set of points in space visited by a Brownian motion.

We now consider some energy spectra of fluid flows whose vorticity is supported by (i.e., has support in) sets whose fractal dimension is known. As a first example, let $\boldsymbol{\xi}(\mathbf{x}, \omega)$ be, with probability one, a smooth random function with support equal to the whole space (in two or three dimensions). At each $\mathbf{x}$, $\boldsymbol{\xi}(\mathbf{x}, \omega)$ is bounded (we omit from now on the qualifier "with probability one") and $\langle \boldsymbol{\xi}^2 \rangle$ is finite. Since $\langle \boldsymbol{\xi}^2 \rangle = \int_0^\infty k^2 E(k) dk$, $E(k)$ decays more rapidly than k^{-3} for k large. We describe this situation somewhat misleadingly by saying $E(k) \sim k^{-\gamma}$, $\gamma = 3^+$ (it is not clear that the spectrum has a "power law" form $k^{-\gamma}$ for large k). We have also shown that, in the limit $\gamma \to 0$, the Kolmogorov spectrum corresponds to a nonsmooth vorticity field $\boldsymbol{\xi}$.

Consider now a vorticity field whose support is a Brownian walk on a lattice of mesh size h. To avoid working with singularities we can "thicken" the random walk by assuming that each of its "legs" (i.e., bonds that belong to it) is in fact a cylinder of radius $\sigma < h/2$. Orient the walk, so that each segment is a vector. Make the circulation constant, and assume $\boldsymbol{\xi}$ is distributed uniformly on the thickened cross-section of each leg. Since div $\boldsymbol{\xi} = 0$, assume that the walk has been closed somewhere far by bending it so that its end touches its beginning. To make the flow homogeneous, assume that we have many such closed loops of vorticity, distributed at random, forming a statistically homogeneous "suspension", sparse enough so that they do not interfere with each other in the calculations below. There is no claim that such a vorticity field can be generated by the equations of fluid mechanics; however, vorticity generates velocity through $\mathbf{u} = K * \boldsymbol{\xi}$ (the Biot-Savart law); the energy is thus defined and has a spectrum.

Pick a sample flow; and calculate $\boldsymbol{\xi}(\mathbf{x})\cdot\boldsymbol{\xi}(\mathbf{x}+\mathbf{r})$. If $\mathbf{x}$ does not belong to a vortex leg, this product is zero. If $\mathbf{x}+\mathbf{r}$ does not belong to a vortex leg, the same is true. If both $\mathbf{x}$ and $\mathbf{x}+\mathbf{r}$ belong to vortex legs the product is non-zero. Average it. If $\mathbf{x}$ and $\mathbf{x}+\mathbf{r}$ belong to different legs $\langle\boldsymbol{\xi}(\mathbf{x})\cdot\boldsymbol{\xi}(\mathbf{x}+\mathbf{r})\rangle = 0$, since the steps in the walk are independent. Thus $\langle\boldsymbol{\xi}(\mathbf{x})\cdot\boldsymbol{\xi}(\mathbf{x}+\mathbf{r})\rangle = 0$ for $\mathbf{r}$ outside a neighborhood $O(h)$ of the origin of $\mathbf{r}$, and is equal to a large constant (tending to infinity when $\sigma \to 0$) in that neighborhood. As $h \to 0$, $\langle\boldsymbol{\xi}(\mathbf{x})\cdot\boldsymbol{\xi}(\mathbf{x}+\mathbf{r})\rangle$ converges to a multiple of a delta function while the walk converges to a Brownian path whose dimension is $D = 2$.

The Fourier transform of $\langle\boldsymbol{\xi}(\mathbf{x})\cdot\boldsymbol{\xi}(\mathbf{x}+\mathbf{r})\rangle$ is a constant. An average over a sphere of radius k gives $Z(k) \sim k^2$; use of $Z(k) = k^2 E(k)$ gives $E(k) =$ constant $= Ck^{-\gamma}$ with C a constant and γ, the exponent, equal to zero. There is no "energy range", so we need not bother with structure functions.

By comparing our two examples, we see that when D has shrunk from 3 to 2, γ has decreased from 3^+ to 0. This is quite reasonable, since as D becomes smaller, the function $\boldsymbol{\xi}$ becomes more singular and its Fourier transform flattens out. Relation (3.2) shows that this is a general property of the Fourier transform, since it shows that as the support of a function shrinks, the support of its transform grows. However, the scaling given by (3.2) does not change power laws [since $(Kk)^{-\gamma} = K^{-\gamma}k^{-\gamma} =$ constant$\cdot k^{-\gamma}$]. This observation about functions and their Fourier transforms is also the mathematical form of the Heisenberg uncertainty principle of quantum mechanics.[6]

However, dimension does not fully characterize a set, and the support does not fully characterize a function. There is no one-to-one relation between γ and D for functions with spectral power laws. All one can say is that the general trend is to have γ smaller when D is smaller, all other things being equal; it is not clear what the last phrase really means; indeed, we shall find some striking exceptions to this trend.

One can show that Brownian motion is self similar (part of it, appropriately blown-up, is identical to the whole). A spectrum of the form $Ck^{-\gamma}$, even for $\gamma = 0$, is also self similar; $C(k/K)^{-\gamma} = C'k^{-\gamma}$ where C' is a new constant. The self-similarity of Brownian motion and of the resulting spectrum are of course related.

3.5. A First Discussion of Intermittency

It is a well-known experimental fact that the vorticity $\boldsymbol{\xi} = \text{curl } \mathbf{u}$ and the energy dissipation $(\nabla \mathbf{u})^2 = \sum_{ij}(\partial_i u_j)^2$ are unevenly distributed in space. A common manifestation of this fact is the presence of wind gusts on stormy days. It has been argued by a number of authors that this

---

[6]H. Dym and H. McKean, *Fourier Series and Integrals*, Academic, 1972.

unevenness, known as intermittency, modifies the Kolmogorov spectrum so that $E(k) \sim k^{-\gamma}$, $\gamma \neq 5/3$. We shall argue in this section that this is unlikely. The point of this section is not merely polemical; the definitions and conjectures that will be introduced in the analysis will turn out to be useful later, and indeed help to explain the more physical derivations of the Kolmogorov spectrum that will appear in following chapters. Since our emphasis will be on vorticity, we emphasize the unevenness of the vorticity distribution. Note however that if the energy dissipation ϵ is unevenly distributed in space, then $\epsilon = \epsilon(\mathbf{x}, \omega)$, and one does not know whether the $\epsilon^{2/3}$ in equation (3.1) should be interpreted as $\langle \epsilon^{2/3} \rangle$ or $\langle \epsilon \rangle^{2/3}$.

Consider a bounded portion $\mathcal{D}$ of physical space, and the integral $\int_{\mathcal{D}} \boldsymbol{\xi}^2 d\mathbf{x}$, the "enstrophy" in $\mathcal{D}$. If $\boldsymbol{\xi}^2$ is unevenly distributed in $\mathcal{D}$, some portions of $\mathcal{D}$ contribute more to this integral than others. Reasons for this intermittency will appear in later chapters. It is hard to see how intermittency can be reconciled with a gaussian distribution for the components of $\boldsymbol{\xi}$, or of any other derivatives of $\mathbf{u}$. Substantial deviations of distributions of $\boldsymbol{\xi}$ from gaussian distributions are indeed observed. This is hardly surprising, since if $\mathbf{u}$ and all of its derivatives were gaussian all their third moments would vanish and there would be no turbulence. Deviations from gaussian behavior are well documented experimentally. Many authors give values of the flatness (defined for a random variable η as $\langle \eta^4 \rangle / \langle \eta^2 \rangle^2$) and the skewness $\left(\langle \eta^3 \rangle / \langle \eta^2 \rangle^{3/2} \right)$ of quantities such as $\partial u / \partial x$, or more generally $\partial^n u / \partial x^n$, derivatives of a component of $\mathbf{u}$. For a gaussian variable, the flatness is always 3 and the skewness 0. For a quantity such as $\partial u / \partial x$ the flatness is above 3 and the skewness less than 0; the skewness presumably measures the energy transfer between wave numbers. One can construct plausible probability distributions that generate the observed values of flatness and skewness.

It has been suggested[7] that intermittency should affect the value of the exponent γ in the Kolmogorov law $E(k) \sim \epsilon^{2/3} k^{-\gamma}$; the dimensional argument would be modified by the introduction of an additional length scale that characterizes intermittency and the cascade argument would be changed through a change of the relevant length and time scales. In particular, γ would be corrected by an expression that contains D, the fractal dimension of the support of the vorticity or the dissipation. We first give a possible definition of D, and then discuss the conjecture.

The Kolmogorov spectrum is self-similar, as discussed above. This creates a strong presumption that turbulent flow in the inertial range is also self-similar, i.e., that a small portion, suitably blown-up, is indistinguishable from a larger portion. If vorticity concentrates on a small set, then it concentrates further on a subset of that set, in a Cantor-set-like sequence

---

[7]B. Mandelbrot, 1977.

of concentrations. As a result, it is likely that the vorticity concentrates on a fractal set. Since a fractal set is not characterized by a single length scale, we understand the difficulty in defining "scale" in general. The scale characterizing intermittency could be the largest distance between disjoint components of the support of vorticity.

Clearly, one cannot expect $\boldsymbol{\xi}$ to die out to exactly zero over a large region in a turbulent flow. One needs to find out not where $\boldsymbol{\xi} \neq 0$ but where $\boldsymbol{\xi}$ is significant. One possible construction goes as follows: Pick a sample flow field (fixed ω). Pick a finite region Λ of space, and look for a subset Λ_ϵ of Λ such that for all ϵ, $\int_{\Lambda_\epsilon} \boldsymbol{\xi}^2 d\mathbf{x} \geq (1 - \epsilon) \int_\Lambda \boldsymbol{\xi}^2 d\mathbf{x}$ (i.e., Λ_ϵ contains almost all of the enstrophy). One would like the "smallest" such Λ_ϵ, but it is not clear how a smallest member of a family of sets is to be picked. Suppose that one can find $D = \inf D_\epsilon$, where D_ϵ is the dimension of Λ_ϵ, and furthermore, that D is independent of ω. Then D is the dimension of the essential support of the vorticity, or, more loosely, the dimension of the set that contains all but a negligible fraction of the vorticity. The interesting case occurs when $D < 3$; this can happen only if $R^{-1} = 0$ or in the limit $R^{-1} \to 0$.

To see that this definition can make sense, consider the equation $\partial_t u + \partial_x(u^2) = 0$. For fixed initial data, this equation does not generate random behavior, therefore we assume the data are random at $t = 0$: $u(x,0) = g(x,\omega)$, $g(x,\omega)$ smooth for each ω. Each flow will develop shocks; if we define $\xi = \partial_x u$, ξ will have the form $\xi(x,\omega) = \phi(x,\omega) + \sum A_j \delta(x - x_j)$, ϕ smooth, A_j random coefficients, $\delta = $ Dirac delta, x_j random variables, and the sum has a finite number of terms that do not vanish in a bounded region. Take $\Lambda = $ segment of length 1; then $\Lambda_\epsilon = \{x_j\}$, x_j in Λ, and for all ω, $D_\epsilon = 0$ (the fractal dimension of a finite set of points is zero). We do not know enough about the Euler and Navier-Stokes equations to say whether this definition makes sense for three-dimensional flow.

The suggestion would be that γ is a function of D, and that if $D \neq 3$ then $\gamma \neq 5/3$. How is this suggestion to be understood? Presumably, the Euler equations and the inviscid limit of the Navier-Stokes equations are each characterized by a unique value of γ and of D (possibly the same pair (γ, D) for the Euler and the Navier-Stokes cases) and a functional relation between two numbers is not a sensible thing to look for. In addition, neither γ (if it exists) nor D come close to uniquely characterizing a flow, so a general functional relation between them in a family of flows larger than Euler/Navier-Stokes flows is not sensible either. (One can however construct interesting one-parameter families of divergence-free vector fields, only one of which corresponds to hydrodynamical flows, in which γ and D both vary and $d\gamma/dD$ can be defined, and we shall do so below.)

The reasonable interpretation of the suggestion that γ depends on D and

$D \neq 3$ lies elsewhere. Turbulent flow contains many scales, has large fluctuations (i.e., the departure of the flow from its mean is not microscopic) and all the scales are strongly coupled. Similar situations are encountered in the theory of "critical phenomena" (see Chapter 6), where the statistics are characterized by "critical exponents" similar to γ. In that other context, one has to be careful to take the fluctuations into account when one calculates the exponents, or else the exponents are incorrectly evaluated. Theories in which fluctuations are ignored are known as "mean-field theories". If one identifies the presence of fluctuations with intermittency, and the Kolmogorov theory with a mean field theory, then one has to correct the Kolmogorov theory by taking D into account.

We shall however obtain $\gamma \sim 5/3$ below from a theory that takes fluctuations into account explicitly, and there is thus no logical need for a correction (although of course one may possibly be needed anyway, since the theory is far from complete). We shall indeed eventually show that the $E(k) \sim k^{-2}$ spectrum derived above by a cascade construction is a better candidate for a "mean-field" result.[8] If that is so, then the change from k^{-2} to $k^{-5/3}$ is an "intermittency correction", and this change has the right sign, since γ typically decreases as D decreases. Intermittency is important, but $\gamma = 5/3$ remains a good guess.

Observe that the definition of D given above, as the dimension of the essential support of the squared vorticity (the enstrophy), greatly oversimplifies the fractal structure that one can plausibly expect in turbulence. If $\boldsymbol{\xi}^2$ is not evenly distributed in space, why should it be evenly distributed in Λ_ϵ? Suppose it is not. One can then have subsets of Λ_ϵ with dimension $D' < D$ that contain a finite fraction of the enstrophy. Define $F(\alpha) = q$, where q is the least upper bound of numbers q' such that

$$\int_{\Lambda_\alpha} \boldsymbol{\xi}^2 d\mathbf{x} \geq q' \int_\Lambda \boldsymbol{\xi}^2 d\mathbf{x}, \quad 0 < \alpha < 3,$$

for some set Λ_α of fractal dimension α. If $\boldsymbol{\xi}^2$ is evenly distributed on Λ_ϵ then each set of dimension $D' < D = \dim \Lambda_\epsilon$, having zero measure in dimension D, contains a zero fraction of $\int \boldsymbol{\xi}^2 d\mathbf{x}$. Thus $F(\alpha) = 0$ for $\alpha < D$, $F(\alpha) = 1$ for $\alpha \geq D$. If $F(\alpha)$ has a different form, one has a "multifractal" vorticity distribution. In this case, subsets of Λ of various dimensions contain nontrivial parts of the vorticity. For a multifractal vorticity distribution, a conjecture that γ depends on dimension must be interpreted as stating that γ is a functional of F.[9]

---

[8] A. Chorin, 1988.

[9] For more references to multifractals, see, e.g., U. Frisch and G. Parisi, 1985, C. Meneveau and K. Sreenivasan, 1987.

As a final remark, note that we have only considered inertial ranges in three space dimensional flow. Cascade arguments and dimensional arguments based on the idea that enstrophy cascades through an inertial range in two space dimension have been offered, and we wait until the next chapter to dismiss them.

4
Equilibrium Flow in Spectral Variables and in Two Space Dimensions

We provide an introduction to equilibrium and non-equilibrium statistical mechanics and apply these theories to truncated spectral approximations and to inviscid flow in two space dimensions. The special feature that makes the two-dimensional theory work is the invariance of the total vorticity $\int \xi d\mathbf{x}$.

4.1. Statistical Equilibrium

Consider[1] a system of N particles, with equal masses m, positions $\mathbf{x}_i = (x_{1i}, x_{2i}, x_{3i})$ and momenta $\mathbf{M}_i = (M_{1i}, M_{2i}, M_{3i})$, $i = 1, \ldots, N$, placed in a box isolated from its surroundings. List the components of the $\mathbf{x}_i$ in order: $x_{11}, x_{21}, x_{31}, x_{21}, \ldots$ and relabel them: $q_1, \ldots, q_{3N}$. Similarly, relabel the momenta $p_1, p_2, \ldots, p_{3N}$. A complete description of the system requires $6N$ numerical values for the q_i, p_i, and can be represented by a single point in a $6N$-dimensional space, the phase space.

Suppose the motion is described by a Hamiltonian H:

$$(4.1) \qquad \frac{dq_i}{dt} = \frac{\partial H}{\partial p_i} \;, \qquad \frac{dp_i}{dt} = -\frac{\partial H}{\partial q_i} \;, \qquad i = 1, \ldots, 3N \;.$$

---

[1] See, e.g., L. Landau and E. Lifshitz, *Statistical Physics*, Pergamon, 1980; D. Chandler *Introduction to Modern Statistical Physics*, Oxford, 1987.

H is constant in time by assumption, and its constant value is the energy. Given initial data, equation (4.1) will describe the subsequent evolution of the system. N is assumed to be large.

The initial data for all the particles in a realistic physical system are unknown, and can be reasonably viewed as random. Consider a probability density f of initial data in phase space, $f = f(\tilde{q}_1, \ldots, \tilde{q}_{3N}, \tilde{p}_1, \ldots, \tilde{p}_{3N}) = f(\tilde{q}, \tilde{p})$, defined as

$$f(\tilde{q}, \tilde{p}) d\tilde{q} d\tilde{p} = P(\tilde{q}_1 < q_1 \leq \tilde{q} + d\tilde{q}_1, \ldots, \tilde{p}_{3N} < p_{3N} \leq \tilde{p}_{3N} + d\tilde{p}_{3N}) ,$$

where P denotes a probability measure and $d\tilde{q} = d\tilde{q}_1, \ldots, d\tilde{q}_{3N}$, etc. $f(\tilde{q}, \tilde{p})$ must obey consistency conditions, as described in Section 2.2. Motion, as described by equations (4.1), will change f (for example, if a particle of velocity 1 has a probability P_1 of being between 0 and 1, after a unit time it will have probability P_1 of being between 1 and 2). We shall omit the tildes when the omission does not lead to ambiguity.

The velocity field that moves the probabilities in phase space has components

$$\left(\frac{dq_1}{dt}, \ldots, \frac{dq_{3N}}{dt}, \frac{dp_1}{dt}, \ldots, \frac{dp_{3N}}{dt} \right) .$$

It satisfies

$$\sum_{i=1}^{3N} \left(\frac{\partial}{\partial q_i} \frac{dq_i}{dt} + \frac{\partial}{\partial p_i} \frac{dp_i}{dt} \right) = \sum \left(\frac{\partial}{\partial q_i} \frac{\partial}{\partial p_i} - \frac{\partial}{\partial p_i} \frac{\partial}{\partial q_i} \right) H = 0 ;$$

by equation (1.1), it is incompressible (this is Liouville's theorem).

The probability density f is invariant, and describes a statistical (or "thermal") equilibrium, if it does not change as the flow in phase space evolves, i.e., if the probability of finding a given set of locations and momenta does not change in time. If one has a set in phase space that is invariant in time (i.e., no systems enter it and no systems leave it), and if one makes f constant in that set, then by incompressibility the resulting probability density is an equilibrium density. Statistical equilibrium is different from a mechanical equilibrium, in which all the p_i are zero.

If $I_1, I_2, \ldots I_M$ are invariants of the flow, i.e., functions of $q = (q_1, \ldots, q_{3N})$ and p such that $\frac{d}{dt} I_j(q, p) = 0$, $j = 1, \ldots, M < 6N$, then the intersection of the sets where $a_j \leq I_j \leq a_j + \Delta a_j$ is invariant, and the constant density on it is an equilibrium density that defines an "equilibrium measure" or "equilibrium ensemble". A typical system in classical mechanics has five, and only five, constants of motion: mass, three components of momentum, and energy. Mass is conserved in any evolution. The momentum can be set to zero by picking coordinates in which the system is at rest (thus the theory we develop will concentrate on the internal motion of the system to the exclusion of its macroscopic motion, i.e., the motion of its container). Then the energy is the only constant in play. (We shall encounter in fluid mechanics systems with more constants.) If we consider the "energy shell" $E \leq H(q,p) \leq E + \Delta E$ and distribute f uniformly on that shell, we obtain the "microcanonical" equilibrium ensemble. It is an axiom that this ensemble describes correctly the equilibria that occur in isolated systems. In the limit $\Delta E \to 0$, one has an equilibrium ensemble that consist of systems distributed uniformly on the energy surface $H = E$.

The entropy $S = S(E)$ of a system is defined as $\log \Lambda$, where Λ is the area of the "energy surface" $H = E$; the temperature T is defined by $T^{-1} = dS/dE$. (In most presentations of statistical mechanics T and S appear with a constant factor k, Boltzmann's constant, whose purpose is to convert units from natural energy units to degrees; we shall omit this factor.)

To see that these definitions make sense, consider the case of an ideal gas, which is a collection of non-interacting point particles. $H = \frac{1}{2m} \sum_{i=1}^{3N} p_i^2$; the radius of the sphere $H = E$ is $\sqrt{2mE}$; its area is

$$\Lambda = C_{3N}(2mE)^{(3N-1)/2} ,$$

where C_{3N} is the appropriate prefactor that depends only on N. Thus

$$S = \log C_{3N} + \frac{3N - 1}{2} \log(2mE)$$

$$\frac{dS}{dE} \cong \frac{3}{2} \frac{N}{E} \qquad \text{for } N \text{ large,}$$

(4.2)
$$T = \frac{2}{3} \frac{E}{N} .$$

One can use the statistical mechanics machinery to calculate the pressure p for this system and derive the perfect gas law $pV = RT$, where V is the volume and $R = N$ in our units, with $N = $ number of particles in a mole of matter. We do not need these results and therefore omit them. Together with the discussion below of convergence to equilibrium, they establish that T is indeed the usual thermodynamical temperature. Equation (4.2) shows that for a perfect gas T is proportional to the energy per particle. Suppose we slowly add particles to this isolated system, without changing the energy. It is hard to imagine how one would do that in a classical mechanical system, but it will turn out below that this is exactly what one does do in a three-dimensional vortex system. The result of such an addition of particles without change of energy is a reduction in temperature. This is an important point.

An equilibrium is an invariant probability measure in phase space. A specific system travels in the regions where the measure is not zero. A system in statistical equilibrium when viewed macroscopically appears to be at rest internally, for the following reason: Many points in phase space correspond to a single observable macroscopic state. For example, the exchange of the positions and momenta of two particles changes the location of the system in phase space but makes absolutely no difference to an observer who looks at the system in physical space. It is generally the case that most, indeed the overwhelmingly largest portion, of the surface $H = E$ corresponds to a single observable system in physical space, and motion on the surface $H = E$ is not detectable macroscopically. The area of $H = E$ is proportional to the number of ways one can combine particles to produce the single visible macroscopic system; the entropy is the logarithm of that number.

One may well wonder how, and indeed whether, equilibrium is reached if one starts away from equilibrium. Suppose we consider a collection of similar systems whose representations in phase space occupies a small neighborhood on $H = E$. If equilibrium is reached, these systems will move to the large portion of $H = E$ that corresponds to the macroscopic equilibrium. Average properties of the system will be computable by averaging over $H = E$. One must however remember that the flow in phase space is incompressible. It is therefore likely that the small initial neighborhood breaks up into thin, long filaments that cover $H = E$, and look evenly distributed as long as one does not look with too fine a microscope. The problem of showing that this indeed happens is difficult and on the whole unsolved, except for very special systems. We shall encounter this difficulty below in the context of vortex dynamics.

When the probability on an invariant portion of phase space is constant, all microscopic states (those described by the $6N$ variables) are equally likely. The probability of occurrence of a macroscopic state (one that can be observed in physical space) is proportional to Λ, and S, the entropy, is the logarithm of a probability, $P = \exp(S)$.

Consider a small subsystem of an isolated system in equilibrium at an energy E; what is the probability that the subsystem has energy E_s, $E_s \ll E$? For the question to be meaningful, one must be able to assign an energy to the subsystem, therefore its coupling to the rest of the full system must be weak enough so that the energy of interactions between the part and the whole can be overlooked. It is not always easy to quantify what the last sentence must means.

The probability $P(E_s)$ that the subsystem has energy E_s equals the probability $\Lambda(E - E_s)$ that the remainder of the system has energy $E - E_s$, where $\Lambda = \Lambda(E - E_s)$ is the area of the surface $H(p, q) = E - E_s$. An expansion in Taylor series gives

$$\log \Lambda(E - E_s) = \log \Lambda(E) - E_s \frac{d \log \Lambda(E)}{dE} + \text{small terms} ,$$
$$\cong \log \Lambda(E) - E_s/T ,$$

and thus

$$P(E_s) \propto e^{-E_s/T} ,$$

where $\propto$ denotes proportionality. Since $\sum_s P(E_s) = 1$, $P(E_s) = Z^{-1} e^{-E_s/T}$, where $Z = \sum_s e^{-E_s/T}$ is the "partition function". More generally, if s is a state of the subsystem, its probability is $P_s = Z^{-1} e^{-E_s/T}$. The summations are over all states, and the sums must be interpreted as integrals when there is a continuum of states. This probability is the canonical, or Gibbs, probability density or "ensemble". For large N, an average with respect to the Gibbs probability density should equal an average with respect to the microcanonical density, since one can average first over subsets of a large set and then over all the subsets. The average $\langle f \rangle$ of a function $f = f(p,q)$ of the state of the system is $\sum f(p,q) P(p,q) = \sum f(p,q) e^{-E(p,q)/T}/Z$ $(= \int (f(p,q) e^{-E(p,q)/T})/Z) dp dq$ when appropriate). In particular, $\langle E \rangle = \sum E_s e^{-E_s/T}/Z$. Writing $\beta = 1/T$, this becomes $\langle E \rangle = (\sum E_s e^{-\beta E_s})/Z = -\frac{\partial}{\partial \beta} \log Z$. If N is large, $\langle E \rangle$ in the canonical ensemble should approximate well the fixed E of the microcanonical ensemble.

We shall need below a formula for the entropy S in the canonical ensemble. Consider the function

$$(4.3) \qquad \tilde{S} = -\sum_s P_s \log P_s$$

where $P_s = e^{-\beta E_s}/Z$ is the probability of the s-th state, and we continue to write sums even when integrals would be appropriate. We have

$$\tilde{S} = -\sum P_s(-\beta E_s - \log Z)$$
$$= \beta \langle E \rangle + \log Z , \qquad (\text{since } \Sigma P_s = 1) .$$

Then

$$\frac{d\tilde{S}}{d\langle E \rangle} = \langle E \rangle \frac{d\beta}{d\langle E \rangle} + \beta + \left(\frac{\partial}{\partial \beta} \log Z \right) \frac{d\beta}{d\langle E \rangle} = \beta = \frac{1}{T}$$

Thus $\tilde{S} = S$, the entropy (up to an immaterial constant). Formula (4.3) yields $\tilde{S} = \log \Lambda$ for the microcanonical ensemble and defines a useful "entropy" even in problems not related to mechanics.

Finally, note that the entropy of disjoint subsets of an isolated system in equilibrium is additive. If Ω_1, Ω_2 are two such subsets, one can view the probability densities connected with them as independent; by the definition of independence, the integral of the product measure defined on their union is the product of the integrals of measures defined on each, and the logarithm of these integrals is thus additive: $S(\Omega_1 \cup \Omega_2) = S(\Omega_1) + S(\Omega_2)$, where S is the entropy.

4.2. The "Absolute" Statistical Equilibrium in Wave Number Space

The theory of the preceding paragraph has been applied to the Euler equation in spectral form[2] and the results are instructive. For the sake of simplicity, consider first the model hyperbolic problem $\partial_t u + \partial_x(u^2) = 0$ in a periodic domain. Expand u in a Fourier series: $u = \sum \hat{u}_k e^{ikx}$. The $\hat{u}_k$ satisfy

$$(4.4) \qquad \frac{d}{dt}\hat{u}_k + ik \sum_{k'} \hat{u}_{k'}\hat{u}_{k-k'} = 0$$

Set to zero all amplitiudes $\hat{u}_k$ with $|k| > K_{\max}$. One can heuristically justify this truncation by assuming that one is trying to solve the corresponding viscous problem $\partial_t u + \partial_x(u^2) = \nu\partial_x^2 u$, where $\nu > 0$ is a viscosity, and that viscosity decreases all the Fourier coefficients that correspond to k large enough. Equation (4.4) now becomes a finite set of ordinary differential equations. Since u is real, $\hat{u}_{-k} = \hat{u}_k^*$. Assume that there is no mean flow, i.e., the average of u over a period is zero, and thus $\hat{u}_0 = L^{-1}\int u\,dx = 0$. One can readily check that the energy E:

$$E = \sum_{k=k_1}^{K_{\max}} \hat{u}_k\hat{u}_k^* \; ,$$

(where k_1 is the smallest non-zero wave number) is a constant of the motion. It is easier to think of real-valued systems, so write $\hat{u}_k = \alpha_k + i\beta_k$, α_k, β_k real, $\alpha_0 = \beta_0 = 0$, with $\hat{u}_{-k} = \alpha_k - i\beta_k$, and $E = \Sigma(\alpha_k^2 + \beta_k^2)$. The solution of (4.4) can be represented as a point in a $2N$-dimensional "phase space" where N is the integer proportional to $K_{\max}$, $K_{\max} = \frac{2\pi}{L}N$, where L is the period of the flow.

---

[2]See e.g. S. Orszag, 1970, and the references therein.

If $\hat{u}_0 = 0$, the sum in equation (4.4) does not include $\hat{u}_k$. Thus $\frac{\partial}{\partial \alpha_k} \frac{d\alpha_k}{dt} = 0$. Similarly, $\frac{\partial}{\partial \beta_k} \frac{d\beta_k}{dt} = 0$, and

$$(4.5) \qquad \sum_{k=k_1}^{K_{\max}} \frac{\partial}{\partial \alpha_k} \frac{d\alpha_k}{dt} + \frac{\partial}{\partial \beta_k} \frac{d\beta_k}{dt} = 0 \ .$$

The flow in phase space of an ensemble of systems that obey (4.4) is incompressible [see equation (1.2) in Section 1.1]. Thus the construction of the preceding section applies. The system has an invariant measure, constant on the sphere $E(\alpha_1, \ldots, \alpha_N, \beta_1, \ldots, \beta_N) = $ constant, known as the "absolute" equilibrium. In that equilibrium, $\langle |\hat{u}_k|^2 \rangle = $ constant $= E/N$.

One hopes that the reader is, at this point, very surprised. The solutions of the equation $\partial_t u + \partial_x(u^2)$ develop shocks, i.e., jump discontinuities, and nothing worse.[3] A smooth solution punctured by jump discontinuities has Fourier coefficients $O(k^{-1})$ for large k, and its energy spectrum for large k is $O(k^{-2})$, not $O(1)$. Either the truncation does something very drastic to the equation, or the equilibrium is never reached, or both.

One can carry out a similar analysis for the Euler equations in two and three space dimensions, equations (1.31) of Section 1.5, with all Fourier coefficients such that $\max(k_1, k_2, k_3) > K_{\max}$ removed. The energy of this finite system of ordinary differential equations is a constant of motion. If $\hat{\mathbf{u}}_0 = 0$ (no mean flow), and $\hat{\mathbf{u}}_{\mathbf{k}} = \boldsymbol{\alpha}_{\mathbf{k}} + i\boldsymbol{\beta}_{\mathbf{k}}$, the equations for $d\boldsymbol{\alpha}_{\mathbf{k}}/dt$, $d\boldsymbol{\beta}_{\mathbf{k}}/dt$ do not have $\boldsymbol{\alpha}_{\mathbf{k}}$ or $\boldsymbol{\beta}_{\mathbf{k}}$ on their right-hand side and (4.5) holds. One then has an "equipartition" ensemble, with $\langle |\hat{\mathbf{u}}_{\mathbf{k}}|^2 \rangle$ independent of $\mathbf{k}$. This "equipartition" ensemble can be produced formally for the untruncated Euler and even the Navier-Stokes equations.[4] The spectrum, which involves an integral over an energy shell in wave number space, is then $E(k) \sim k$ in two space dimensions, $E(k) \sim k^2$ in three, for $k < K_{\max}$. The Kolmogorov spectrum has certainly not been recovered.

What has gone wrong? The fluid flows we are interested in have certain smoothness properties, since they are described by the Euler and Navier-Stokes equations. These properties are not taken into account in the truncated system, and as a result, sample flows with $\langle |\hat{\mathbf{u}}_{\mathbf{k}}|^2 \rangle = $ constant are very unsmooth. Fluid flows also have a number of integral invariants ($\int \xi^k d\mathbf{x}$ in two space dimensions, circulation around arbitrary contours in inviscid three-dimensional flow) which are destroyed by the truncation. We have seen in the previous section that invariants play a major role in the theory of statistical equilibria, and we shall see below that every step towards taking smoothness and invariance properties into account will produce spectra

---

[3]See e.g. P. Lax, *Hyperbolic Conservation Laws and the Mathematical Theory of Shock Waves*, *SIAM*, 1972.

[4]E. Hopf, 1952.

closer to what is reasonable. One should not draw the conclusion, from the "absolute" equilibrium result, that the Kolmogorov spectrum is a manifestation of non-equilibrium. On the other hand, we shall see that under appropriate circumstances, when certain invariants become unimportant, the k^2 spectrum will reappear.

4.3. The Combinatorial Method: The Approach to Equilibrium and Negative Temperatures

We now rederive the main results of Section 4.1 by a method that, though elementary, will be of use later and will in particular allow a discussion of the relaxation to statistical equilibrium.[5]

We specialize the discussion to the Hamiltonian system of two-dimensional vortex dynamics. The phase space is $2N$- rather than $6N$-dimensional for N particles; $q_1, \ldots, q_N$ denote the x_1 coordinates of the particles; and $p_1, \ldots, p_N$ denote the x_2-coordinates. The translation of our results to the usual particle system is immediate. Consider the plane in which the vortices move, and divide it into M boxes of side h and area h^2, $M \ll N$. Divide the N particles among the M boxes, assuming that there are particles n_i in the i-th box, n_i large for all i. One can think of each box as having reached statistical equilibrium while the system as a whole has not reached equilibrium.

The partition of the N particles among the boxes produces a well-defined macroscopic state. What is its probability? In other words, what volume in the $2N$-dimensional phase space corresponds to this macroscopic state? Clearly, this volume is proportional to $h^{2n_1} \cdots h^{2n_2} = h^{2N}$. It is also proportional to the number of ways N objects can be divided into M subsets of sizes $n_1, \ldots, n_M$, since an exchange of particles inside the boxes produces a new point in phase space but does not alter the macroscopic picture. The probability of a given partition $n_1, \ldots, n_M$ is then

$$W = \left(\frac{N!}{n_1! \cdots n_M!} \right) h^{2N} ,$$

when the term in parantheses is the number of ways of dividing N objects into M groups of n_1, n_2, etc.; it is assumed that $n_i \gg 1$ for all i, $M \ll N$. We define $S = \log W$ to be the entropy of our (not necessarily equilibrium) system.

Suppose one can identify an energy $n_i E_i$ connected with the i-th box, with $E = \Sigma n_i E_i$ the total energy. For a vortex system, in which the interactions are long range, this is a questionable assumption, and we shall

---

[5]See, e.g., A. Sommerfeld, *Thermodynamics and Statistical Mechanics*, Academic Press, 1964.

eventually abandon it. It is made here so that the argument parallels the argument that led to the canonical ensemble and so that the conclusions can be compared with the canonical ensemble. Thus $\Sigma n_i E_i = E$, $\Sigma n_i = N$.

We now wish to maximize W or S subject to these constraints; the result will be the most probable partition and it should be a statistical equilibrium. For large n, $n! \cong (\frac{n}{e})^n$ (Stirling's formula) and thus

(4.6) $$S = \log W = \text{constant} - \Sigma n_i \log n_i$$

subject to the constraints $\Sigma n_i E_i = E$ and $\Sigma n_i = N$. Using Lagrange multipliers, we find at the maximum

$$n_i = e^{-\alpha} e^{-\beta E_i} \; ,$$

with α, β Lagrange multipliers.

We wish to compare this formula with the canonical ensemble. The equilibrium of the box system is obtained from (4.3) by calculating the probability P_s of every partition s of the N vortices among the M boxes, and summing $-\Sigma P_s \log P_s$. If, however, the vortices are thrown among the boxes independently of each other, one can check that this sum reduces to $-N\Sigma P_i \log P_i$, where P_i is the probability that one vortex lands in the i-th box. In the case of independent throws, an application of Tchebysheff's theorem yields $n_i/N \sim P_i$, and thus $S = -n_i \log n_i$ is the entropy. The condition $\Sigma n_i = N$ yields $n_i = \frac{N}{Z} e^{-\beta E_i}$, $Z = \sum_i e^{-\beta E_i}$, and a comparison with the canonical ensemble yields $\beta = 1/T$, $T = $ temperature.

The assumption of independent throws is plausible when N is large, because then the constraints that link the n_i are not strong. The fact that we recover the canonical distribution can be taken as evidence that the assumption is acceptable.

One advantage of this rederivation is that it allows us to define the entropy of a system not in equilibrium, and to assert a fundamental principle: the entropy of an isolated system never decreases. At equilibrium, for such a system, the entropy is maximum. The derivation also reemphasizes the role of the constants of motion in equilibrium theory: any additional constraint, for example, the existence of a property Q_i attached to the i-th box and subject to the constraint $\Sigma n_i Q_i = Q$, Q a constant of the motion of the underlying time-dependent system, produces a new Lagrange multiplier, say γ, and modifies the partition to $n_i = NZ^{-1} e^{-\beta E_i - \gamma Q_i}$, $Z = \Sigma e^{-\beta E_i - \gamma Q_i}$.

The temperature T still satisfies $T^{-1} = \frac{dS}{dE}$ ($E = \langle E \rangle$), assuming S depends only on E. We have seen that for the ideal gas $\frac{dS}{dE} > 0$; the same holds for the absolute equilibrium of Section 4.2 and in fact for most systems one is used to, but this inequality is not a law of nature. One can perfectly well imagine systems such that for E moderate there are many ways of arranging their components so that the energy adds up to E but for E large there are only a few ways of doing so. Then derivative dS/dE is negative for E large enough and T is negative. This situation will indeed occur for vortex systems. If $T \cdot > 0$, low-energy states have a high probability, and if $T < 0$, high-energy states have a high probability.

Suppose one takes two systems, each separately in equilibrium, one with energy E_1 and entropy S_1, the other with energy E_2 and entropy S_2. Suppose one joins them; the resulting union has energy $E_1 + E_2$ and is not necessarily in equilibrium. Its entropy, initially $S = S_1 + S_2$, will increase in time. Then

$$\frac{dS}{dt} = \frac{dS_1}{dt} + \frac{dS_2}{dt} = \frac{dS_1}{dE_1}\frac{dE_1}{dt} + \frac{dS_2}{dE_2}\frac{dE_2}{dt} > 0 \ ,$$

while energy is conserved:

$$\frac{dE_1}{dt} + \frac{dE_2}{dt} = 0 \ .$$

Therefore

$$\frac{dS}{dt} = \left(\frac{dS_1}{dE_1} - \frac{dS_2}{dE_2} \right) \frac{dE_1}{dt} = \left(\frac{1}{T_1} - \frac{1}{T_2} \right) \frac{dE_1}{dt} \ .$$

Suppose $T_2 > T_1$, both positive; then $\frac{dE_1}{dt} > 0$, i.e., energy moves from the hotter body to the colder body. Now suppose $T_2 < 0$. It still follows that $\frac{dE_1}{dt} > 0$, i.e., a body with negative temperature is "hotter" than a body with positive temperature. Negative temperatures are above $T = \infty$, rather than below absolute zero. Further, the canonical formulas show that $T = -\infty$ is indistinguishable from $T = +\infty$; $|T| = \infty$ is the boundary between $T < 0$ and $T > 0$. In terms of $\beta = T^{-1}$, temperature increases as β varies from infinity to zero through positive values, and then from zero to minus infinity through negative values.

Consider a physical system in equilibrium, with $T < 0$. Divide it into M subsets, and endow each subset with a macroscopic velocity U_i, $i = 1, \ldots, M$. The energy of the system is thus $E = \frac{1}{2}m_i U_i^2 + \Sigma E_i$, where m_i is the mass of the i-th box and E_i is the internal energy in the i-th box, $E_i = \langle E_i \rangle$ in earlier notations. The sum of the momenta $\Sigma m_i U_i$ must be zero if the system as a whole does not move in its coordinate system. The entropy S_i in the i-th box is a function of E_i only, and if $dS_i/dE_i = T^{-1} < 0$ the entropy increases when energy moves from the E_i to the U_i. Thus a system at $T < 0$ should be expected to have large scale motion even at equilibrium.[6]

---

[6]L. Landau and E. Lifshitz, *Statistical Mechanics*, Pergamon, 1980; note however the difference in the conclusions.

4.4. The Onsager Theory and the Joyce-Montgomery Equation

Consider a collection of N vortices of small support occupying a finite portion $\mathcal{D}$ of the plane, of area $A = |\mathcal{D}|$. The area can be made finite by surrounding it with rigid boundaries, in which case the vortex Hamiltonian must be modified through the addition of immaterial smooth terms; alternatively, one can confine the vortices to a finite area initially and conclude that they will remain in a finite area, because the center of vorticity $\mathbf{X} = \Sigma\Gamma_i\mathbf{x}_i/\Sigma\Gamma_i$, $\mathbf{x}_i =$ positions of the vortices, and the angular momentum $\Sigma\Gamma_i^2|\mathbf{x}_i - \mathbf{X}|^2$ are invariant.[7] For the moment, we only consider inviscid flow with all the $\Gamma_i = 1$.

The entropy of this system is

$$S = -\int_{\mathcal{D}^N} f(\mathbf{x}_1,\ldots,\mathbf{x}_N)\log f(\mathbf{x}_1,\ldots,\mathbf{x}_N)d\mathbf{x}_1,d\mathbf{x}_2\ldots d\mathbf{x}_N \ ,$$

where f is the probability that the first vortex is in a small neighborhood of $\mathbf{x}_1$, the second in a small neighborhood of $\mathbf{x}_2$, etc. (See the expression for the entropy in a canonical ensemble in Section 4.1.) The energy of this system is $E = H + B$, where H is the two-dimensional vortex Hamiltonian and B is an appropriate constant. The entropy is maximum when

$$(4.7) \qquad\qquad f = \text{constant} = A^{-N} \ ,$$

as can readily be seen from the combinatorial argument without an energy constraint. The corresponding energy is

$$(4.8) \qquad \langle E_c \rangle = -\frac{1}{4\pi}N(N-1)\int_{\mathcal{D}} d\mathbf{x}\int_{\mathcal{D}} d\mathbf{x}'\log|\mathbf{x}-\mathbf{x}'| + B$$

on the average. Clearly, one can produce a larger E by bunching vortices together, and thus if S has no local maxima other than (4.6), $T^{-1} = dS/d\langle E \rangle < 0$ for $\langle E \rangle > \langle E_c \rangle$. This is Onsager's observation.[8] If $T > 0$, the Gibbs factor $\exp(-E/T)$ gives a high probability to low-energy states, and if $T < 0$, high-energy states are favored; the latter are produced by bunching together vortices, forming large, concentrated vortex structures. The $f = $ constant state is the $|T| = \infty$ boundary between $T < 0$ and $T > 0$. The T introduced here has no connection whatsoever with the molecular temperature of the underlying fluid; as pointed out in Section 1.1, the molecular degrees of freedom and the vortex variables are insulated from each other.

To give this argument a more quantitative form, we return to the combinatorial method of the last section.[9] We assume there are N vortices

---

[7]G. Eyink and H. Spohn, 1992.

[8]L. Onsager, 1949; the presentation of Onsager's result follows the previous reference.

[9]G. Joyce and D. Montgomery, 1973.

(the limit $N \to \infty$ will be considered in the next section). N^+ vortices have $\Gamma = 1$, N^- have $\Gamma = -1$, $N^+ + N^- = N$, a slight generalization of the earlier combinatorial argument. We divide $\mathcal{D}$ into M boxes, with n_i^+ positive and n_i^- negative vortices in each. The corresponding W is

$$W = \left(\frac{N^+!}{n_1^+! \cdots n_M^+!} \right) \left(\frac{N^-!}{n_1^-! \cdots n_M^-!} \right) h^{2N} .$$

The entropy is $S = \log W$ (remember the assumption of independent throws), which is to be maximized subject to the constraints $\Sigma n_i^+ = N^+$, $\Sigma n_i^- = N^-$, and

$$E = \tfrac{1}{2} \sum_i \sum_{j \neq i} (n_i^+ - n_i^-) G_{ij}(n_j^+ - n_j^-) = \text{constant} ,$$

where $G_{ij} = -\frac{1}{2\pi} \log |\mathbf{x}_i - \mathbf{x}_j| + C$, $\mathbf{x}_i$ is in the i-th box, $\mathbf{x}_j$ is in the j-th box, and C is a constant chosen so that $E \geq 0$. This E approximates the energy of a vortex system; we have abandoned the assumption $\Sigma n_i E_i = E$ used in the last section to bring out the canonical distribution; it must now be assumed that the vortex system is part of a larger isolated system, with the remainder of the system acting as a "heat bath", i.e., a source of interactions that allow E to vary with T fixed, but do not change the area available to the vortex system or the total vorticity. The maximization of S leads to the equation

$$\log n_i^+ + \alpha^+ + \beta \sum_j G_{ij}(n_j^+ - n_j^-) = 0 ,$$

$$\log n_i^- + \alpha^- - \beta \sum_j G_{ij}(n_j^+ - n_j^-) = 0 ,$$

where $\alpha^+, \alpha^-, \beta$ are Lagrange multipliers. A little algebra yields

$$
\begin{aligned}
n_i^+ - n_i^- &= \exp(-\alpha^+ - \beta \sum_j G_{ij}(n_j^+ - n_j^-)) \\
&\quad - \exp(-\alpha^- + \beta \sum_j G_{ij}(n_j^+ - n_j^-)),
\end{aligned}
$$

(4.9)

for $i = 1, \ldots, M$. Let $h \to 0$ so that $n_i^+ - n_i^- \to \xi(\mathbf{x})h^2 = \xi(\mathbf{x})dx_1 dx_2$, $(\exp(-\alpha^+))/h^2 \to d^+$, $(\exp(-\alpha^-))/h^2 \to d^-$, and $\Sigma G_{ij}(n_i^+ - n_i^-) \to \int G(\mathbf{x} - \mathbf{x}')\xi(\mathbf{x}')d\mathbf{x}'$, where $G(\mathbf{x}) = -\frac{1}{2\pi} \log |\mathbf{x}| + C$. One can easily check that this limit is self-consistent. Equation (4.9) converges to

$$\xi(\mathbf{x}) = d_+ \exp(-\beta \int G(\mathbf{x} - \mathbf{x}')\xi(\mathbf{x}')d\mathbf{x}')$$

(4.10)
$$+ d_- \exp(\beta \int G(\mathbf{x} - \mathbf{x}')\xi(\mathbf{x}')d\mathbf{x}')$$

where d_+, d_- are appropriate normalization coefficients.

Let ψ be the stream function, $u_1 = -\partial_2 \psi$, $u_2 = \partial_1 \psi$; an easy calculation gives $\Delta\psi = -\xi$, $\Delta = $ Laplace operator and $\psi = -\int G(\mathbf{x} - \mathbf{x}')\xi(\mathbf{x}')d\mathbf{x}'$. Thus,

(4.11)
$$-\Delta\psi = \xi(\mathbf{x}) = d_+ e^{-\beta\psi} - d_- e^{\beta\psi} .$$

This is the Joyce-Montgomery equation. In a periodic domain one can set $\psi = 0$ on the boundary of the domain, and $N^+ = N^- = N/2$, $d_+ = d_- = d$. Then

$$2d = \frac{N}{\int d\mathbf{x} e^{\beta\psi}} ,$$

$$-\Delta\psi(\mathbf{x}) = \xi(\mathbf{x}) = d \sinh \beta\psi(\mathbf{x}) .$$

If $N^+ = N$, $N^- = 0$, then $d_- = 0$, $d^+ = N/Z$, $Z = \int_D e^{-\beta\psi} d\mathbf{x}$, and

(4.12)
$$-\Delta\psi = \xi(\mathbf{x}) = \frac{N}{Z} \exp(-\beta\psi(\mathbf{x})) .$$

In either case, ξ is a function of ψ. The Euler equation is

$$\partial_t \xi = -u_1\partial_1\xi - u_2\partial_2\xi$$
$$= (\partial_2\psi)(\partial_1\xi) - (\partial_1\psi)(\partial_2\xi) = J(\psi, \xi) ,$$

where $J = $ Jacobian of ξ, ψ, which is zero when $\xi = \xi(\psi)$. The resulting average flow is a stationary (time-independent) solution of the Euler equation, with macroscopic motion, as expected when $\beta < 0$.

The appearance of ψ should not be surprising. We know from Chapter 1 that in two dimensions ψ and H are closely related. Equation (4.11) and (4.12) are vortex versions of the canonical distribution. One obtains a partial differential equation in two variables x_1, x_2 because (i) the vortex Hamiltoniain involves on x_1, x_2 as conjugate variables, and one does not have to worry about the distribution of particles in boxes in a four-dimensional position/momentum space, (ii) the stream function and the Hamiltonian are related, (iii) the vorticity is proportional to the one-particle probability density, and (iv) the independent throws approximation has been used.

It should be emphasized at this point that the ξ we have calculated is not only a specific solution of Euler's equation, but more importantly it is the stationary average density of the vorticity. Specific flows may depart from this average, but we expect the departure to be small. The reasons can be found in the references given earlier in this section; the theory we have given is a "mean field theory" with small fluctuations, as can be shown to be appropriate.

For $\beta \geq 0$ and for $-8\pi N < \beta < 0$ equation (4.12) can be shown to have solutions. In the latter case the solutions are non-unique; the solutions have multiple peaks; the solution that maximizes the entropy has a single sharp but smooth peak. Equation (4.11) has a double peak when the entropy is maximum and $-16\pi N < \beta < 0$, one positive peak and one negative peak.

For $\beta < -8\pi N$ (i.e., "hotter" than $T = -1/8\pi N$), the Joyce-Montgomery equation with $\xi \geq 0$ has no classical solution and in fact does not describe reasonable physics. To understand why, consider a system of two vortices with $\Gamma = +1$ in a bounded region $\mathcal{D}$.[10] The canonical distribution applies to small systems as well as to large ones, and therefore applies here. Fix one vortex, and consider all possible positions of the other. If the separation of the vortices is r, the energy of the pair is $E = -\frac{1}{4\pi}\log r$; the Gibbs factor $\exp(-E/T)$ is $\exp(+(\beta/4\pi)\log r) = r^{\beta/4\pi}$; the partition function is $Z = \int_{\mathcal{D}} r\,dr\,d\theta\; r^{\beta/4\pi} = \text{constant} \times r^{(2+\beta/4\pi)}$. If $\beta < -8\pi$, Z blows up as $r \to 0$; almost all of the contribution to $Z = \sum_s P_s$ will come from a single state where $r \cong 0$, ($r = 0$ if $\phi_\delta = \delta$), and the system will collapse to a point. Thus, after repeating the argument with N vortices, one expects (4.12) to have only distribution valued solutions of no physical significance when $\beta < -8\pi N$, as is indeed the case.[11] Similar arguments hold for equation (4.11); the collapse of an N vortex system with $|\Gamma_i| \neq 1$ will be discussed in the next section.

4.5. The Continuum Limit and the Role of Invariants

Consider what happens to the theory of the previous section as the number N of vortices tends to infinity, while the system remains confined to a finite region $\mathcal{D}$. We present a mostly descriptive discussion of this question, with proofs available in the references. The conclusion is that one can reformulate the equations so that the limit $N \to \infty$ is meaningful.

One can readily conclude from formulas (4.7) and (4.8) of Section 4.4 that when $|T| = \infty$, the energy E is proportional to N^2 for large N, and S is proportional to N. One expects that as N increases, the limits

$$\lim_{N \to \infty} \frac{E}{N^2} = e(\mathcal{D}) \;,$$

---

[10] R. Kraichnan and D. Montgomery, 1980.
[11] E. Caglioti et al., 1992.

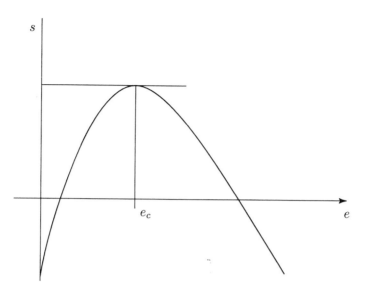

FIGURE 4.1. Scaled entropy and energy.

$$\lim_{N \to \infty} \frac{S}{N} = s(\mathcal{D})$$

exist. Thus $E = N^2 e$, $S = Ns$. If the strength of the vortices is $\Gamma \neq 1$, $E = N^2 \Gamma^2 e$; we consider for simplicity the case $\Gamma_i = \Gamma > 0$ for all i. One expects $s(e)$ to be the maximum of the entropy over all probability density functions which produce the given scaled energy e. All these expectations are fulfilled.[12] $s = s(e)$ can be calculated, in particular numerically, and looks as in Figure 4.1.

The asymptotic slope on the right is -8π. Note that e is growing when T is, if one allows for the peculiar fact that $T < 0$ is hotter than $T > 0$. One can define a "scaled" temperature $\tilde{T}$ by $\tilde{T}^{-1} = \frac{ds}{de}$, $\tilde{T} = \tilde{T}(e)$. Then $e = e(\tilde{T})$, $s = s(\tilde{T})$. The usual temperature $T^{-1} = \frac{dS}{dE} = Nds/(N^2\Gamma^2 de)$ satisfies $T = N\Gamma^2\tilde{T}$; if one chooses N and Γ so that $N\Gamma = \xi_0 = $ constant, keeping the vortex density constant as $N \to \infty$, then $T = (\xi_0^2/N)\tilde{T}$ and $|T| \to 0$ as $N \to \infty$. The collapse of the Joyce-Montgomery equation occurs when $\tilde{\beta} = \tilde{T}^{-1} = -8\pi$. Thus T is a function of N, but the physics remain the same if $\tilde{T}$ is constant. The sign of T is the sign of $\tilde{T}$ and depends on $s(e)$ only; it is determined by the distribution in $\mathcal{D}$ of the normalized vorticity $\xi(\mathbf{x})/\int_{\mathcal{D}} \xi(\mathbf{x})d\mathbf{x}$.

Suppose $\xi_0 = 1$, so that $N\Gamma = 1$, $\beta = N\tilde{\beta}$. A rederivation of the Joyce-Montgomery equation in the case of positive vorticity, that takes into account the dependence of the various quantities on N, yields

---

[12]R. Robert, 1991; J. Miller, 1991; G. Eyink and H. Spohn, 1992.

$$-\Delta\tilde{\psi} = \frac{e^{-\tilde{\beta}\tilde{\psi}}}{\int_D e^{-\tilde{\beta}\tilde{\psi}}d\mathbf{x}} = \frac{1}{\tilde{Z}}e^{-\tilde{\beta}\tilde{\psi}} \; ,$$

where $\tilde{\psi}$ is $O(1)$ as $N \to \infty$. There is an analogous equation for the two-sign case. This equation can be solved analytically in the special case $\mathcal{D}$ = circle of radius 1 with $\mathbf{u} \cdot \mathbf{n} = 0$ on $\partial\mathcal{D}$.[13] Indeed, if one assumes radial symmetry, $\tilde{\psi} = \tilde{\psi}(r)$, $r = |\mathbf{x}|$; thus

$$\frac{1}{r}\frac{d}{dr}\left(r\frac{d}{dr}\tilde{\psi}\right) = -\frac{1}{\tilde{Z}}e^{-\tilde{\beta}\tilde{\psi}} \; ; \qquad \tilde{\psi}(1) = 0 \; .$$

changing variables, so that $q = \log r$, $H(q) = -\tilde{\beta}\tilde{\psi}(e^q) + 2q$, we find

$$\frac{d^2}{dq^2}H = \tilde{\beta}\frac{1}{\tilde{Z}}e^H \; ;$$

setting $y = e^H$, we obtain

$$y(q) = \frac{4EA\tilde{Z}}{\tilde{\beta}}e^{\sqrt{2E}q}(1 - Ae^{\sqrt{2E}q})^{-2} \; ,$$

where $E = \frac{1}{2}\left(\frac{dH}{dq}\right)^2 - \frac{\tilde{\beta}}{\tilde{Z}}e^H$ is the constant energy and A is a constant of integration. For $\tilde{\beta} > -8\pi$ we then find, using the boundary condition,

$$A = \frac{\tilde{\beta}}{8\pi + \tilde{\beta}} \; , \qquad \tilde{Z} = \pi(1 - A) \; , \qquad E = 2 \; ,$$

and

(4.13)
$$\tilde{\xi} = -\Delta\tilde{\psi} = \frac{1 - A}{\pi}\frac{1}{(1 - Ar^2)^2} \; .$$

As expected, the solution exists as a smooth function only for $\tilde{\beta} > -8\pi$. It has a sharp peak at the origin.

It is interesting to contrast these results with the properties of the "absolute equilibrium" of Section 4.2. That "absolute equilibrium" in spectral variables is at a temperature T proportional to E/K_{max}^3, where E is the energy and T is always positive. The calculation follows exactly the steps in the calculation of the temperature of an ideal gas and need not be repeated. As the number of variables K_{max}^3 tends to infinity, this temperature tends to zero through positive values. Remember that the $T > 0$ and $T < 0$ ranges touch when $|T| = \infty$, thus if $\tilde{T} < 0$, the temperatures of the "absolute" equilibrium and of the vortex equilibrium diverge from each other.

---

[13]E. Caglioti et al., 1992.

The spectral representation distorts reality because it fails to impose the conservation of $I_1 = \int \xi(\mathbf{x})d\mathbf{x}$ that is automatically imposed in a vortex representation. This constant is responsible for the reduction in the volume of phase space available to the system as the energy increases, and thus creates the negative temperatures. The spectral form of this constraint is not easy to write down and impose when the spectral representation is truncated.

It is natural to ask whether one does not have to impose all the conservation laws $I_k = \int \xi^k d\mathbf{x} = \text{constant}$ in order to obtain the correct statistics. This has not been done so far for $k > 1$; if $\phi_\delta = \delta$ in the vortex representation (point vortices) the second and higher powers of ξ are not defined; if ϕ_δ is smooth, the supports of the $\phi_\delta(\mathbf{x} - \mathbf{x}_i)$ may overlap to a variable extent and the I_k, $k > 1$, may vary. An elegant construction that imposes all the constraints has been carried out by Miller[14]: Divide $\mathcal{D}$ into small squares of area h^2, and make ξ constant on each. Consider only the states that can be obtained from a single state by permuting the squares, no two squares ever coming to rest on top of each ("self-avoiding configurations", in the language of Chapter 6 below). Clearly all the I_k are the same in all these configurations. The log of the number of distinct configurations for a given E is the entropy. One can construct the whole theory in terms of these permutations; the results are unchanged in their main features. Thus the constancy of I_1 is enough, but less is not.

The canonical ensemble has been consistently used in this section. This raises questions of principle. The simple derivation of the canonical ensemble from the microcanonical ensemble in Section 4.1 does not apply to vortex systems, as can be seen by contrasting the treatment of the energy constraint in Sections 4.1 and 4.3. If the vortex system is viewed as being in equilibrium with an external "heat bath" rather than with the remainder of an isolated vortex system, what kind of heat bath has a negative temperature, how does it interact with the vortex system, and is its temperature T or $\tilde{T}$? One could argue that only the microcanonical ensemble makes sense for vortex systems, and thus one has to establish the equivalence of the ensembles for vortices. This has been partly done,[15] but serious questions remain. One can encounter situations with $T < 0$ where the two ensembles may fail to be equivalent.[16]

A comparison of this chapter with Section 3.1 shows that in two space dimensions the universal equilibrium postulated there is simply statistical equilibrium. The independence of small and large scales reduces to the statement that small scales are asymptotically insignificant.

---

[14] J. Miller, 1991.

[15] See e.g. G. Eyink and H. Spohn, 1992.

[16] M. Kiessling, private communication, 1992.

The two-dimensional statistical theory is well verified by numerical experiment (Figure 4.1). It has been suggested that certain large-scale vortical structures seen in two-dimensional flows, for example, Jupiter's Great Red Spot, are examples of vortex equilibria in nature.[17]

4.6. The Approach to Equilibrium, Viscosity, and Inertial Power Laws

Statistical equilibria are of interest only if they are reached from most initial data. There is strong evidence, mainly numerical, that the two-dimensional equilibria constructed above are in fact reached. Some general statements can be made about the relaxation to equilibrium, and some questions remain open.

Suppose one starts from initial data that consist of two patches of vorticity, say $\xi = 1$ in sets C_1, C_2, both bounded, C_1, C_2 disjoint, and $\xi = 0$ elsewhere. Since vorticity is merely transported by the fluid motion, one has to imagine a process by which the vorticity in the patches is redistributed so as to match ξ_∞, the solution of the (one-sign) Joyce-Montgomery equation (4.12). (See Figure 4.2.) One can imagine that the boundaries of C_1, C_2 sprout filaments, as in the convergence of subsets of the constant energy surface to the microcanonical ensemble (Section 4.1). The resulting filaments could reorganize so as to approximate ξ_∞ on a sufficiently crude scale.[18]

The filamentation of the boundary should lower the energy. Indeed, if a small vortex patch is broken into two halves that pulled apart, the energy goes down; two vortices of strength $\Gamma = 1$ each, near each other, act as one vortex of strength 2, whose energy is four times that of one of them; two vortices of strength 1 far from each other have an energy that is the sum of their individual energies. To make up for the loss of energy in filamentation the two patches have to approach each other. This process of simultaneous filamentation and consolidation is well documented numerically[19] (Figure 4.3). Similarly, one expects a non-circular patch to become nearly circular with a halo of filaments, the whole approximating ξ_∞ on a rough scale. Even a circular patch with non-constant ξ, increasing from its center outward, can reorganize its vorticity so that filaments shoot off while energy is being conserved. On the other hand, a patch with ξ decreasing as one moves away from the center is presumably stable, and belongs to the set of initial data that do not approach ξ_∞; such a patch of course does in itself constitute a rough version of ξ_∞.

---

[17]P. Marcus, 1988.

[18]See, e.g., D. Dritschel, 1988.

[19]See, e.g., A. Chorin, 1969; T. Buttke, 1990.

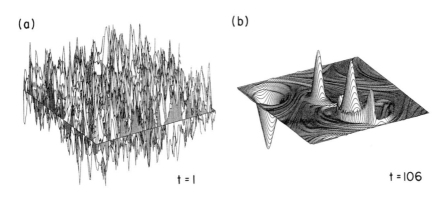

(a) t = 1

(b) t = 106

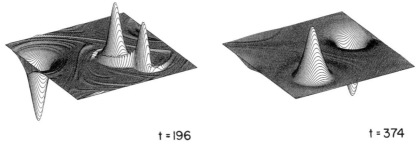

(c) t = 196

(d) t = 374

FIGURE 4.2. Convergence of ξ to the solution of the two-sign Joyce–Montgomery equation as time increases. [Reprinted with permission from Montgomery et al., *Phys. Fluids A*, **4**, 3–6 (1992).]

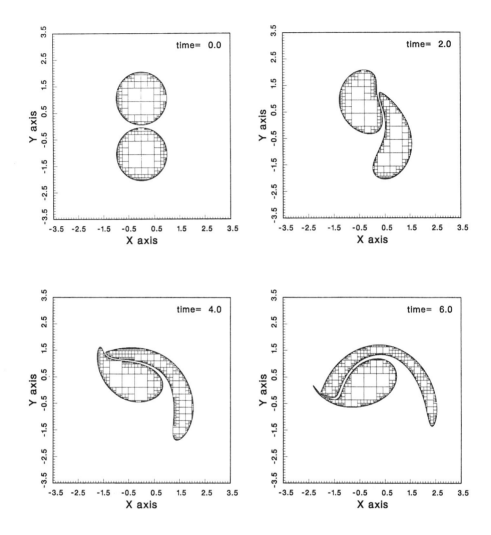

FIGURE 4.3. A consolidation/filamentation event. (Reprinted with permission from T. Buttke, *Journal of Computational Physics* **89**, 161–186 (1990).)

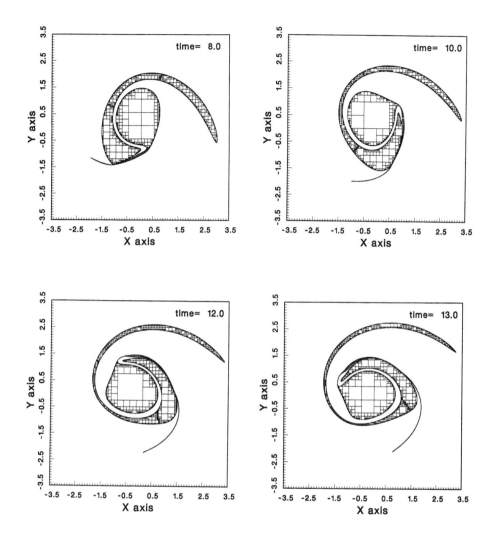

FIGURE 4.3. *Continued.*

This process of simultaneous filamentation and consolidation can be deduced from the invariance of the energy and the enstrophy in spectral form: $\int E(k)dk = $ constant, $\int k^2 E(k)dk = $ constant. If some energy moves towards the large k's (small scales), then even more energy must move towards the small k's (large scales). On the whole, there is an energy "cascade" toward the small k's.

If the initial ξ is complicated, and has many maxima and minima, one can imagine, and indeed see on the computer, a process of progressive curdling, in which nearly circular patches that look locally like ξ_∞ first form on small scales, then slowly migrate towards each other and consolidate if viewed on a crude enough scale. The curdles can never truly merge, since the flow map ϕ_t is one-to-one. At each stage of this curdling the nearly circular patches are nearly independent, with whatever correlations their locations have manifesting itself only on large scales. The flow can then be approximated as $\sum \eta_i \xi_\infty (\mathbf{x} - \mathbf{x}_i)$, $\eta_i = $ random coefficients. The energy spectrum is approximately proportional to $|\mathbf{k}|^2 |\hat{\xi}_\infty (\mathbf{k})|^2$, where $\hat{\xi}_\infty$ is the Fourier transform of $\xi_\infty (\mathbf{x})$, and is a property of each curd individually (see Section 2.4).[20] One then has local equilibria slowly consolidating into larger equilibria, as postulated in the combinatorial analysis of Section 4.3. At each time t the curdling process can be described by a length scale L (for example, the mean distance between curds), and one can imagine that $L \sim t^\alpha$ with α a universal exponent.[21] One can also argue that the "absolute equilibrium" provides a reasonable description of the curdling system as long as $K_{\max} < 1/L$.[22]

This successive curdling picture provides a suggestion as to what happens in the presence of shear or in complex geometries. In three space dimensions the "universal" aspects of turbulence appear on small scales, and one can readily imagine that arbitrary large-scale structures have "universal" small-scale features. Here, in two dimensions, the universal structures grow to large scales, and an imposed shear or an imposed boundary mass interferes with them. It is readily imagined however that the curdling process will simply stop when it ceases to be compatible with the conditions imposed on the problem.

One can wonder about the effect of a small viscosity ν on the processes just described. To the extent that the effect of viscosity is to smear the small scales, and as long as the time it takes to reach equilibrium is small compared to the time scale of viscous decay, the picture above should be unaffected. One could say a little more: suppose the effect of viscosity is approximated by Brownian motion (Section 2.5). The Brownian motion

---

[20] See also A. Chorin, *Lectures on Turbulence Theory*, 1975, Berkeley Mathematics Dept. Lecture Note Series.

[21] See, e.g., J. McWilliams, 1984.

[22] A. Chorin, 1974.

can be thought of as being generated by the bombardment of the vortices by the molecules of an ambient fluid at a temperature ν. The statement in Section 1.1 that the molecular temperature of underlying fluid has no impact on the velocity field $\mathbf{u}$ is not belied by this analogy; there is no reason to believe that the molecular temperature is ν, and the ambient fluid just introduced is merely notional. The effect of the bombardment that has just been imagined is to couple weakly the notional fluid at the temperature ν with the vortex system, and if $\nu < T =$ vortex temperature, to reduce the latter. If $T < 0$, the cooling of the vortex system brings one closer to the $|T| = \infty$ equidistribution solution, in agreement with the intuitive idea that random pushes should interfere with the formation of concentrated vortices. After a long enough time one would end up with $\xi =$ constant.

It has been argued that two-dimensional flow has inertial ranges with "power law" spectra of the form $E(k) \sim k^{-\gamma}$, with $3 \leq \gamma \leq 4$. If the inertial range is defined as the range above the dissipation range where there are universal (here, equilibrium) phenomena, we find that in two dimensions the inertial range and the energy range coincide, the spectrum is determined by ξ_{∞} which is smooth, and there is no power law. On scales small enough so that the small-scale structure is not necessarily negligible there may be a power law, but it may fail to extend to $k = \infty$ if the solution of Euler's equations is smooth enough. The energy in two dimensions moves from small to large scales, but one can well imagine that $Z(k)$, for example, concentrates around k_1 while its dissipation concentrates around k_2, $k_2 \gg k_1$, when R^{-1} is small but not zero. One can then produce a dimensional argument of the Kolmogorov type for $Z(k)$ and obtain a k^{-3} power law, which is not quite compatible with $\int E(k)k^2 dk =$ constant. If one believes in cascade arguments, one can imagine that enstrophy "cascades" across the range where this power law holds.

The main evidence for this kind of power law comes from numerical calculations, which have their own reasons for producing power laws, and from meteorological observations, which encompass phenomena more complex than mere incompressible flow. The arguments for power laws in two dimensions do not appear to be decisive.[23]

---

[23] See A. Chorin, 1975, *loc. cit.*

5
Vortex Stretching

Vortex motion in three-dimensional space differs from vortex motion in two dimensions in several ways; the most important result from vortex stretching and the consequent non-conservation of vorticity and enstrophy. In this chapter we discuss the salient qualitative aspects of vortex stretching, in preparation for the statistical description in Chapter 7.

5.1. Vortex Lines Stretch

Consider[1] a fluid occupying the whole three-dimensional space; let $\phi_t = \phi_t(\omega)$ be the random flow map induced by an incompressible random flow field; let $\mathbf{x}^0$, $\mathbf{x}^0 + \delta\mathbf{x}^0$ be two nearby points at time $t = 0$; the flow map will map them on $\mathbf{x}^t$, $\mathbf{x}^t + \delta\mathbf{x}^t$ at time t, where

$$\delta\mathbf{x}^t = \phi_t(\mathbf{x}^0 + \delta\mathbf{x}^0) - \phi_t(\mathbf{x}^0) \cong A\delta\mathbf{x}^0,$$

to leading order in $\delta\mathbf{x}^0$. A is a random 3×3 matrix. The length $|\delta\mathbf{x}^t|$ satisfies

$$0 \le |\delta\mathbf{x}^t|^2 = (\delta\mathbf{x}^t, \delta\mathbf{x}^t) \cong (A\delta\mathbf{x}^0, A\delta\mathbf{x}^0) = (\delta\mathbf{x}^0, W\delta\mathbf{x}^0)$$

---

[1] W. Cocke, 1969; S. Orszag, 1970; A. Chorin, 1991b.

where (,) denotes the ordinary inner product and $W = A^T A$ is a positive definite symmetric matrix, with real eigenvalues $\lambda_1, \lambda_2, \lambda_3$ and corresponding normalized orthogonal eigenvectors $\mathbf{e}_1, \mathbf{e}_2, \mathbf{e}_3$. The flow being incompressible, the Jacobian of the flow map is 1 (Section 1.1) and thus

$$|W| = |A| \cdot |A^T| = 1 = \lambda_1 \lambda_2 \lambda_3,$$

where the vertical bars denote a determinant.

At $t = 0$, consider a family of vectors $\delta \mathbf{x}^0$ of a fixed length $|\delta x^0| = \delta x^0$, with directions uniformly distributed over the unit sphere. Denote by $\langle \ \rangle_s$ an average over the unit sphere. Thus,

$$\delta \mathbf{x}^0 = \sum_{i=1}^{3} \delta x_i^0 \mathbf{e}_i,$$

$$|\delta \mathbf{x}^0|^2 = \sum_{i=1}^{3} (\delta x_i^0)^2,$$

and $\langle (\delta x_i^0)^2 \rangle_s = \frac{1}{3}(\delta x^0)^2$, $i = 1, 2, 3$.

For each $\omega \in \Omega$ and each $\delta \mathbf{x}^0$,

$$|\delta \mathbf{x}^t|^2 = (\delta \mathbf{x}^0, W \delta \mathbf{x}^0) = \sum \lambda_i(\omega)(\delta x_i^0)^2,$$

$$\langle |\delta \mathbf{x}^t|^2 \rangle = \left\langle \sum \lambda_i (\delta x_i^0)^2 \right\rangle$$

(ordinary average, $\delta \mathbf{x}^0$ fixed),

$$= \sum_{i=1}^{3} \langle \lambda_i \rangle (\delta x_i^0)^2.$$

Averaging over initial data with equidistributed directions, we obtain

$$\langle \langle |\delta \mathbf{x}^t|^2 \rangle \rangle_s = \left(\frac{1}{3} \sum \langle \lambda_i \rangle \right) (\delta x^0)^2.$$

For any non-negative numbers $\lambda_1, \lambda_2, \lambda_3$,

$$\frac{1}{3}(\lambda_1 + \lambda_2 + \lambda_3) \geq (\lambda_1 \lambda_2 \lambda_3)^{1/3};$$

hence

$$\frac{1}{3} \sum \langle \lambda_i \rangle \geq 1,$$

with equality holding only if all the $\lambda_i = 1$ with probability one—a trivial random flow. Therefore $\langle \langle |\delta \mathbf{x}^t|^2 \rangle \rangle_s$ is increasing,

$$\frac{d}{dt} \langle \langle |\delta \mathbf{x}^t|^2 \rangle \rangle_s > 0,$$

except for trivial turbulence in which the derivative is zero. On the average, over the probability space and over all initial directions, line length increases. Lines in some directions may well shrink while others lengthen; if the flow is isotropic as well as homogeneous, there are no privileged directions and

$$(5.1) \qquad \frac{d}{dt}\langle|\delta\mathbf{x}^t|^2\rangle > 0 \ .$$

Vortex lines are special lines, and constitute a negligible fraction of all lines (there is one vortex direction at each point, but an infinite number of others). All arguments that involve averages with respect to a probability measure may fail to hold in a negligible fraction of cases, and thus one cannot conclude from (5.1) that vortex lines stretch, even in isotropic flow. This conclusion is however eminently plausible.

The rate at which lines stretch is not estimated in this argument, and it may well be small. The well-known numerical experience is that vortex lines stretch very rapidly (and it has been argued that they may stretch infinitely in a finite time).[2] It may well be however that the argument above, which uses only the randomness of the map and incompressibility and does not distinguish between late times and early times, is the best general argument one can give. If the flow is smooth at $t = 0$, its spectrum decays rapidly with k. To form an inertial range, energy must be transferred to the large k range. Vortex stretching is the main mechanism for such transfer: as vortex lines stretch, their cross-sections become smaller, and thus variations of ξ on smaller scales appear. More importantly, we shall see that vortex stretching is accompanied by vortex folding, which is an even more powerful agent of energy transfer. It is plausible that the formation of an inertial range is accompanied by rapid stretching. Once an inertial range has been formed and Kolmogorov scaling holds for both the second- and third-order structure functions, it may be that the rate of stretching decreases (in agreement with the "bathtub" picture).

The argument just given makes no use of the Euler/Navier-Stokes equations, except for the constraint of incompressibility. It is independent of the direction of time: one still gets stretching if $\phi_t(\omega)$ is replaced by $\phi_{-t}(\omega)$. It is the randomness of ϕ_t that is essential. At $t = 0$ $\delta\mathbf{x}^0$ is known, at a later time $\delta\mathbf{x}^t$ is less well known, and as a result $|\delta\mathbf{x}^t|$, appropriately averaged, increases. Remember that the entropy S is given by $S = -\sum_s P_s \log P_s$, where P_s is the probability of the state s. When $t = 0$ there is a single state with $P = 1$, and $S = 0$. At $t > 0$, the randomness of ϕ_t creates a range of possible states and $S > 0$. Thus entropy increases and as a result line length increases. We shall eventually show a converse: line length is a measure of entropy, and an increase in line length produces an increase in entropy. The difficulty in reconciling an irreversible increase in line length

---

[2]See, e.g., A. Chorin, 1982; R. Grauer and T. Sideris, 1991; J. Bell and D. Marcus, 1991.

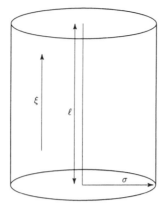

FIGURE 5.1. A vortex cylinder I.

by the reversible Euler equations is analogous to the well-known difficulties in reconciling an irreversible increase in entropy with the reversible
equations of classical mechanics.

Consider a circular cylinder whose base has radius σ, whose height is ℓ,
and containing a vorticity vector $\boldsymbol{\xi}$ parallel to the axis (Figure 5.1). The
enstrophy in that cylinder is $Z = \int_I |\boldsymbol{\xi}|^2 d\mathbf{x} = \pi\sigma^2\ell|\boldsymbol{\xi}|^2$. Suppose the cylin-
der is stretched by a factor α. The height becomes $\ell\alpha$, the radius becomes
$\sigma/\sqrt{\alpha}$ (by conservation of volume), $|\boldsymbol{\xi}|$ becomes $\alpha|\boldsymbol{\xi}|$ (by conservation of
circulation), and thus $Z \rightarrow \alpha^2 Z$. Z increases as a result of stretching (note
that in this example, Z per unit length of the cylinder also increases). It
is generally believed that the average enstrophy increases as a result of
stretching in a homogeneous random flow. An example is not a proof, but
a general proof does not seem to be available; such a proof would require
a lower bound on Z in terms of line length or line complexity.[3]

Assuming $\frac{d}{dt}\langle\boldsymbol{\xi}^2\rangle > 0$, we could use $\langle\boldsymbol{\xi}^2\rangle$ as an "entropy" (not the entropy), i.e., a monotonically non-decreasing function of time that is time
invariant only when the flow is statistically time invariant (= stationary).
Such "entropy" could be used instead of the true entropy, which in three
space dimensions does not have the relatively simple expressions that result
from the peculiarities of the two-dimensional vortex Hamiltonian.

5.2. Vortex Filaments

We want to approximate the vorticity in three space dimensions as a sum
of simple objects, as we have done in the plane. It will be important to
respect the identity $\operatorname{div}\boldsymbol{\xi} = 0$, and to keep in sight the fact that the integral

---

[3]See, e.g., M. Freedman et al., 1992.

lines of $\boldsymbol{\xi}$ can span substantial distances. We shall approximate $\boldsymbol{\xi}$ as a sum $\boldsymbol{\xi} = \sum_i \boldsymbol{\xi}_i(\mathbf{x})$, where supp $\boldsymbol{\xi}_i$ is a closed vortex tube. Any smooth vorticity field can be so represented.[4] However, we shall sometimes go beyond what can be rigorously justified by looking at flow fields in which the union of the supports of the vortex tubes is much smaller than the space available (a "sparse suspension of vortex tubes"). We shall often assume that the cross-sections of the tubes are small compared to their lengths, and that the eccentricity of their cross-section (the ratio of their maximum diameter to their minimum diameter) is bounded. One can argue the following points:

(i) Such vortex tubes are the natural generalizations of the vortices used in two dimensions. In two dimensions one could have used objects of different geometries (for example, vortex sheets) but no loss has been incurred by not doing so.

(ii) Numerical experiment[5] shows that turbulent flow is dominated by vortex tubes of small cross-section and bounded eccentricity.

(iii) Other candidates for generic structures, such as vortex sheets[6], are liable to roll-up instability that creates tubular objects.

(iv) The use of vortex tubes is self-consistent; it will be shown that an ensemble of vortex tubes will remain an ensemble of vortex tubes under reasonable circumstances.

However, the main argument for approximating $\boldsymbol{\xi}$ by a union of vortex tubes is convenience. In order to analyze what happens one must simplify the geometric complexity of turbulence. If the results depend on the sparseness of the vortex suspension, one can try to extend them to "dense" suspensions by a continuation method, as will be explained. One has to pay attention to a possible distinction between intermittency and the sparseness of the vortex suspension. The first is a property of the fluid flow, and the latter may be an artifact of the method of analysis. The thin vortex tubes we shall work with will be referred to as vortex filaments. After we analyze the structure of the vortex cross-sections, it will turn out that the filaments can be so complicated that their characterization as filaments is often a metaphor more than a description.

For a sparse suspension of vortex filaments the invariance of the helicity $\mathcal{H} = \int \boldsymbol{\xi} \cdot \mathbf{u} d\mathbf{x}$ has a particularly simple interpretation.[7] Consider for simplicity two vortex filaments V_1, V_2 with circulation κ_1, κ_2 (note the change of notation). Let C_1, C_2 be the centerlines of V_1, V_2, and let Γ_1 be the

---

[4]See, e.g., T. Beale and A. Majda, 1982b; C. Greengard, 1986.

[5]See, e.g., A. Chorin, 1969; Z.S. She et al., 1990, 1991.

[6]See, e.g., R. Prasad and K. Sreenivasan, 1989.

[7]H. Moffatt, 1969.

circulation around C_1:

$$\Gamma_1 = \int_{C_1} \mathbf{u} \cdot d\mathbf{s}.$$

Similarly for Γ_2. If V_1 is unknotted (i.e., can be spanned by a non-self-intersecting surface Σ),

$$\Gamma_1 = \int_\Sigma \boldsymbol{\xi} \cdot d\Sigma = \begin{cases} 0 & \text{if } V_1, V_2 \text{ are unlinked} \\ \pm\kappa_2 & \text{is } V_1, V_2 \text{ are simply linked} \end{cases}$$

and more generally, $\Gamma_1 = \pm n\kappa_2$ if V_2 turns around V_1 n times. The sign depends on the direction of $\boldsymbol{\xi}$ in V_2. n is the winding number of V_2 around V_1. Γ_1 is an invariant of the flow, and so is $\Gamma_2 = \int_{C_2} \mathbf{u} \cdot ds$. More generally, if there are n unknotted filaments $V_1, V_2, \ldots, V_n$, the quantities $\Gamma_i = \int_{C_i} \mathbf{u} \cdot ds = \sum_j \alpha_{ij}\kappa_j$ are invariants, where C_i is the center-line of the j-th filament and the α_{ij} are integers. The quantity $\Gamma_i\kappa_i$ (no summation) is also an invariant, since each factor is. We have

$$\kappa_i\Gamma_i = \int_{C_i} \kappa_i\mathbf{u} \cdot d\mathbf{s} = \int_{V_i} \mathbf{u} \cdot \boldsymbol{\xi}d\mathbf{x}.$$

Therefore $\int_{\text{space}} \mathbf{u} \cdot \boldsymbol{\xi}d\mathbf{x} = \sum_i \kappa_i\Gamma_i$; the invariance of the helicity is deduced from the constancy of the winding numbers.

If the vortex filaments are knotted, they can be unknotted by a clever insertion of counter-rotating vortex segments that break up vortex loops into smaller loops, and the same conclusion holds. In the presence of viscosity, however small, vortex filaments can reconnect[8]; nearby counterrotating vortices flatten and merge. The transformation of Figure 5.2 is then possible. The approximate conservation of helicity in slightly viscous flow puts constraints on possible reconnection geometries.

5.3. Self-Energy and the Folding of Vortex Filaments

If vortex filaments stretch, they must also fold.[9] The reasons can be seen on an example: suppose a vortex loop as in Figure 1.1 is expanded so that its radius increases from R to αR, $\alpha > 1$. The radius of the arms is reduced from ρ to $\rho/\sqrt{\alpha}$, $|\boldsymbol{\xi}|$ in the arms increases to $\alpha|\boldsymbol{\xi}|$, and the energy $E = \frac{1}{2}\mathbf{M} \cdot \mathbf{u}$ ($\mathbf{M}$ = magnetization, see Section 1.4) increases from $E \sim (\pi\rho^2|\boldsymbol{\xi}|^2)(\log R/\rho)\pi R^2 \cong \pi^2\rho^2|\boldsymbol{\xi}|^2 R$ to $\sim \pi^2\rho^2\alpha^2 R|\boldsymbol{\xi}|^2 = \alpha^2 E$. If energy is conserved, the loop must change its shape so that the velocity fields produced by the vorticity in its several parts cancel to a large extent. We now proceed to give this observation a more quantitative form.

---

[8]See, e.g., C. Anderson and C. Greengard, 1989.

[9]A. Chorin, 1988a.

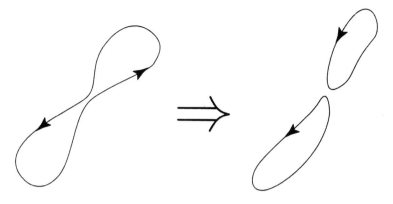

FIGURE 5.2. Vortex reconnection.

Consider again the vorticity in the vortex tube I of Figure 5.1, and consider the energy associated with the vorticity in I, i.e., the integral

$$E_I = \int_I d\mathbf{x} \int_I d\mathbf{x}' \, \frac{\boldsymbol{\xi}(\mathbf{x}) \cdot \boldsymbol{\xi}(\mathbf{x}')}{|\mathbf{x} - \mathbf{x}'|},$$

where $\boldsymbol{\xi}$ is parallel to the axis, $\boldsymbol{\xi} = (0, 0, \xi)$, and $\xi = \xi_0(x_1/\sigma, x_2/\sigma)$, with $\xi = 0$ when $\sqrt{x_1^2 + x_2^2} > \sigma$. E_I depends on σ, ℓ, ξ_0 and the circulation (integral of ξ_0 on the base). We wish to examine how E_I changes when σ, ℓ change; $E_I = E_I(\sigma, \ell)$, with the circulation kept fixed and ξ keeping the form ξ_0. Such changes in E_I can be induced by stretching the cylinder or by slicing it normally to its axis.

Suppose all the coordinate axes are stretched by a factor α. $\mathbf{x} \to \alpha \mathbf{x}$, $d\mathbf{x} \to \alpha^3 d\mathbf{x}$, $d\mathbf{x}' \to \alpha^3 d\mathbf{x}'$, $\xi \to \alpha^{-2}\xi$ (by conservation of circulation), $|\mathbf{x} - \mathbf{x}'| \to \alpha|\mathbf{x} - \mathbf{x}'|$, and thus $E_I(\alpha\sigma, \alpha\ell) = \alpha E_I(\sigma, \ell)$. Pick $\alpha = \sigma^{-1}$, then $E_I(1, \ell/\sigma) = \sigma^{-1} E_I(\sigma, \ell)$, or $E_I(\sigma, \ell) = \sigma \tilde{E}(\ell/\sigma)$, where $\tilde{E}(\ell/\sigma) = \tilde{E}(q)$ is a function of a single argument q whose precise form depends on ξ_0. Clearly, $\frac{d\tilde{E}(q)}{dq} > 0$ (lengthening the tube with σ fixed increases the energy). For q small, $\tilde{E}(q)$ is proportional to q^2 (putting two shallow vortices on top of each other roughly doubles the velocity field and quadruples the energy). For q large, $\tilde{E}(q)/q$ is an increasing function of q (adding up the velocity fields of two long vortices put on top of each other does more than add up their energies, but maybe not by much). An asymptotic analysis that we omit shows that for large q, $\tilde{E}(q)$ is proportional to $|q \log q|$.

Consider now a closed vortex filament V of unit circulation and a small, approximately circular cross-section of small but positive area. Suppose V can be approximately covered by N circular cylinders I_i, $i = 1, \ldots, N$, of equal lengths ℓ and of radii σ_i, $i = 1, \ldots, N$. The energy of the velocity field associated with this filament is

$$
\begin{aligned}
E &= \frac{1}{8\pi} \int d\mathbf{x} \int d\mathbf{x}' \frac{\boldsymbol{\xi}(\mathbf{x}) \cdot \boldsymbol{\xi}(\mathbf{x}')}{|\mathbf{x} - \mathbf{x}'|} \\
&= \frac{1}{8\pi} \sum_i \sum_{j \neq i} \int_{I_i} d\mathbf{x} \int_{I_j} d\mathbf{x}' \frac{\boldsymbol{\xi}(\mathbf{x}) \cdot \boldsymbol{\xi}(\mathbf{x}')}{|\mathbf{x} - \mathbf{x}'|} \\
&\quad + \frac{1}{8\pi} \sum_i \int_{I_i} d\mathbf{x} \int_{I_i} d\mathbf{x}' \frac{\boldsymbol{\xi}(\mathbf{x}) \cdot \boldsymbol{\xi}(\mathbf{x}')}{|\mathbf{x} - \mathbf{x}'|} \\
&= \sum_i \sum_{j \neq i} E_{ij} + \sum_i E_{ii} .
\end{aligned}
$$

(5.2)

Let $\mathbf{t}_i$ be a vector lying along the axis of I_i, originating at the center of I_i, and pointing in the direction of $\boldsymbol{\xi}$ in I_i. If I_i and I_j are far from each other,

$$
E_{ij} \sim \frac{1}{8\pi} \frac{\mathbf{t}_i \cdot \mathbf{t}_j}{|i - j|},
$$

where $|i - j|$ is the straight-line distance between I_i and I_j (say, between their centers). The error made by assuming that this formula holds whenever $i \neq j$ is not large, because most of the distances $|i - j|$ are large compared to ℓ, except if the filament V is very folded, which is what we want to show. E_{ii} is a function of ℓ and σ_i, which we assume to be the same function of σ_i and $\ell/\sigma_i = q$ for all i. (For the limits of this assumption, see Section 5.4). Thus $E_{ii} = \sigma_i \tilde{E}(\ell/\sigma_i)$, with $\frac{dE_{ii}}{d\sigma_i} < 0$.

Suppose V is stretched (by the ambient velocity field and by its own velocity field), in such a way that the shape of its cross-section does not change radically. Suppose further that the stretched vortex filament can be approximated by $N' > N$ cylinders of length ℓ. The sum $\sum_i E_{ii}$ associated with the new filament will be larger because it will have more elements and most of them will be larger, their cross-sections having shrunk. The energy of the new filament will be

$$
\frac{1}{8\pi} \sum_{i=1}^{N'} \sum_{\substack{j=1 \\ j \neq i}}^{N'} \frac{\mathbf{t}_i \cdot \mathbf{t}_j}{|i - j|} + \sum_{i=1}^{N'} E_{ii}(\text{new}).
$$

The double sum has now $O(N'^2)$ entries, and has decreased. The typical $|i - j|$ cannot increase if the filament remains connected, and thus the inner products $\mathbf{t}_i \cdot \mathbf{t}_j$ must decrease, i.e., the filament must fold, or else the energy of the vortex increases.

In inhomogeneous flow, for example, in the presence of shear, the energy can increase, and thus vortex stretching can act as an energy sink for turbulent flow. The stretching and folding transfer energy to smaller scales.

In equation (5.2), the double sum is the "interaction energy" and the sum $\sum E_{ii}$ is the "self-energy". The term analogous to the self-energy was omitted when the two-dimensional vortex Hamiltonian was derived from the energy of two-dimensional flow because it was constant, albeit possibly infinite. Here it cannot be omitted without discussion because it can vary. The division of the energy into self-energy and interaction energy is arbitrary, as it depends on the choice of cylinder length ℓ; the smaller ℓ, the larger the share of the interaction energy.

Suppose there is a natural length ℓ in the problem (for example, suppose the curvature of vortex lines is known to be bounded by ℓ^{-1}). Suppose the cross-sections of the tubes are comparable; the energy of the vortex can be written as

$$(5.3) \qquad E = \sum_i \sum_{j \neq i} E_{ij} + \mu' N \ , \qquad \mu' = E_{ii} \quad \text{for any} \ \ i.$$

The quantity $\mu = \frac{\partial E}{\partial N}$ is the "chemical potential" of a vortex segment; the name comes from the applications of this quantity in chemistry. As is obvious from the discussion above, $\mu = \frac{\partial E}{\partial N}$ and μ' are defined here qualitatively. $\mu = \frac{\partial E}{\partial N} = \sum_{j \neq N} E_{Nj} + \mu'$. If the vortex filament is smooth, E_{Nj} decays rapidly as $|j - N|$ increases, and $\frac{\partial E}{\partial N}$ is $O(\mu')$. If the vortex filament is well folded, μ and μ' are likely to be very different.

5.4. Fractalization and Capacity

As vortex filaments stretch and fold, their axes converge to fractal sets. It is immaterial for our purposes here whether this happens in a finite time, so that the fractal limits are actually achieved, or in an infinite time, so that the fractals appear only asymptotically. In this section we consider only the axes of the filaments. What happens to the cross-sections will be discussed later.

Consider a bounded subset C of the plane, and consider all the vorticity functions ξ such that $\xi \geq 0$, $\int \xi d\mathbf{x} = 1$, supp $\xi \subset C$ (i.e., all ways of distributing a unit of vorticity among the points of C). For each ξ, consider the energy of the resulting flow, including the self-energy of any small pieces of C. Given C, is it possible to make the energy in a finite volume containing C finite? If the answer is yes, C is said to have positive capacity, and if the answer is no, C is said to have zero capacity.[10]

A set consisting of a single point has zero capacity because a point vortex has an infinite self-energy. A set consisting of a finite number of points also has zero capacity. A set that has finite area has positive capacity. Thus a

---

[10] O. Frostman, 1935.

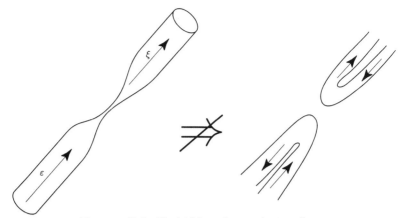

FIGURE 5.3. Forbidden change in topology.

large enough set has positive capacity. It turns out that any set in the plane that has positive Hausdorff dimension has positive capacity. The problem of finding planar sets which can support a unit vorticity field with finite energy is identical to the problem of finding planar sets that can support a unit electrical charge in such a way that the resulting two-dimensional electrical energy is finite.

The generalization of the two-dimensional electrical problem to three dimensions is straightforward: a bounded set in three dimensions has positive scalar capacity if it can support a unit charge with a finite electrical energy. It is not necessary to use a finite box surrounding the set, because the energy decays fast enough as one moves towards infinity so that the energy outside a large enough box is bounded, as we have seen in Chapter 1. A set of Hausdorff dimension larger than one has positive capacity, and a set of Hausdorff dimension less than one has zero capacity. The case of Hausdorff dimension equal to one requires further analysis. Generalization to the case of vorticity in three dimensions is much more difficult, because $\boldsymbol{\xi}$ is a vector and must satisfy the constraint $\operatorname{div}\boldsymbol{\xi} = 0$.

A set in three-dimensional space has positive vector capacity if it can be the support of a vorticity field $\boldsymbol{\xi}$ such that $\operatorname{div}\boldsymbol{\xi} = 0$, $\int_{\Sigma}\boldsymbol{\xi}\cdot d\Sigma = 1$ for at least one smooth surface Σ, and such that the resulting velocity field has a finite energy. Otherwise, the set has zero vector capacity. For any $\epsilon > 0$, it is possible to find a set of Hausdorff dimension $1 + \epsilon$ which has positive vector capacity: consider a planar set C_1 of positive Hausdorff dimension ϵ contained in a bounded set, place on it a unit planar vorticity which has finite planar energy, construct a vertical cylinder of base C_1, with vorticity vector parallel to its axis, and bend that cylinder into a large vortex ring.

A smooth closed curve has dimension one and zero vector capacity. Consider a vortex filament that is stretched without changing the topology of

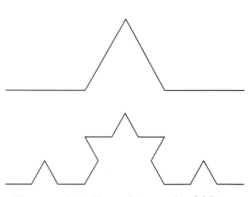

FIGURE 5.4. Fractalization by folding.

its vorticity field, i.e., the transformation in Figure 5.3 is not allowed. If any portion of the filament is stretched so that it collapses onto a curve, that curve carries vorticity. The associated energy is infinite. If the energy is conserved in the stretching, this kind of collapse is forbidden. Thus stretching must be accompanied by a folding that prevents the appearance of smooth vortex filaments of zero cross-section. Note that smooth curves can be stretched into objects of fractal dimension larger than one without ceasing to have topological dimension one (i.e., being deformable to a smooth curve). As an example,[11] consider the straight line and the sequence of stretchings in Figure 5.4: The middle third of the segment is stretched up, then the middle third of the four new segments is stretched up, etc. Assume without proof that fractal dimension and Hausdorff dimension are identical in this case. The end result of this sequence of operations consists of four pieces, each similar to the whole with a similarity ratio $1/3$, and as in the Cantor set calculation, the end result has dimension $D = \log 4/\log 3 > 1$. The length of the resulting curve is infinite (each operation multiplies it by $4/3$). It is of course not obvious that when this particular curve carries vorticity the energy is finite. Fractalization is necessary but not sufficient for keeping the energy finite.

5.5. Intermittency

Consider a vortex filament that intersects a box of finite volume, and suppose the filament stretches and the box moves so that it continues to be crossed by the filament. The portion of the filament in the box lengthens as it folds, or else the forbidden collapse to a curve will occur. One may wonder whether folds uniformly spread in the box are sufficient to conserve

---

[11]B. Mandelbrot, 1975, *loc. cit.*

energy, or whether the filament must fold into ever tighter folds, in which pairs of counter-rotating filaments are confined in an ever shrinking portion of the available volume. In the latter case, the origin of intermittency is explained. If D' is the fractal dimension of the volume into which the vortex folds must shoehorn themselves, then D' is a bound on D, the dimension of the essential support of the vorticity. The following heuristic argument[12] suggests that $D' < 3$.

Energy appears on ever smaller scales as a result of stretching and folding. Suppose some specific scale has been reached. Smaller scales will be produced if a portion of the existing filament is stretched, say by a factor α. To simplify matters, assume α is constant along the stretching portion of the tube. If $\alpha = 1$ the stretching has stopped. If $\alpha > 1$, the next stretching will add another factor of α, and the filament will fractalize, as in the examples of the preceding section. The radius σ of the filament will decrease by a factor $1/\sqrt{\alpha}$ when the filament stretches by α.

Write the energy in the form

$$E = E_1 + \mu' N,$$

where $E_1 = \sum_i \sum_{j \neq i} E_{ij}$ is the interaction energy, and $\mu' N$ is a shorthand for $\sum_i E_{ii}$, the self-energy. Compare $\tilde{E}(\sigma/\sqrt{\alpha})$, the interaction energy of the part stretched by a factor α, with $E_1 = E_1(\sigma)$, its energy before stretching. $\tilde{E}(\sigma/\sqrt{\alpha})$ is of order $(\alpha^2/a)E(\sigma)$, where a is the ratio of the old to the new average distance between points on the filament. Indeed, the filament is α times longer and $|\mathbf{x}-\mathbf{x}'|$ is a times larger after stretching. Very distant pieces contribute little to $\tilde{E}$, and it is reasonable to assume a priori that $a \sim \alpha$ and $\tilde{E}$ increases by α. When one considers the self-energy $\mu' N$, one sees that the stretched portion consists of $\alpha\sqrt{\alpha}$ pieces, each $1/\sqrt{\alpha}$ times smaller in all directions than before, and thus each having a self-energy $1/\sqrt{\alpha}$ smaller than before; $\mu' N$ thus increases by α also. The sum of the two energies should be finite, and thus it makes sense to assume E_1 is negative, and the two increments in the energy cancel each other. However, under the assumptions just stated, this will not happen. If the stretching is smooth, the radius of curvature of the filament increases proportionally to α, and thus $O(\sqrt{\alpha})$ of the pieces that contribute to E_1 are aligned, and make a positive contribution $O(\alpha)$ to E_1. To keep the energy finite one must assume that $a < \alpha$ or that the radius of curvature of the filament increases more slowly than α, and therefore the filament folds into tighter bundles than what is dictated by the available volume. Note that the non-unique division of E into E_1 and $\mu' N$ is merely a convenient but not necessary device for characterizing coherent contributions to E.

---

[12] A. Chorin, 1988b.

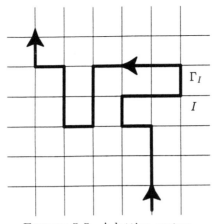

FIGURE 5.5. A lattice vortex.

To make this argument more precise, imagine a three-dimensional cubic lattice, with bond length h, and a "vortex filament" that coincides with a connected sequence of bonds in that lattice. The quotation marks will be dropped in the future. As before, there is no claim that such a vortex filament can be generated by the equations of motion; the lattice is a useful device for estimating the effects of stretching and folding, to the extent that they do not depend on the precise form of the equations of motion; its usefulness comes from the possibility of enumerating configurations and keeping track of possible singularities in the energy integral.

The vortex, and hence the occupied bonds (the "vortex legs") must be oriented consistently, so that circulation is constant along the filament. The bonds can be thickened so that each vortex leg has a radius $\sigma \leq h/2$. Each vortex site (intersection of bonds) can lie on at most two bonds, or else $\boldsymbol{\xi}$ is not single valued. The energy of the vortex is $E = \sum_I \sum_{J \neq I} \boldsymbol{\Gamma}_I \cdot \boldsymbol{\Gamma}_J / |I - J| + N\mu'$, where $I = (i_1 h, i_2 h, i_3 h)$ is a multi-index denoting a lattice site, $\Gamma_I = |\boldsymbol{\Gamma}_I|$ is the circulation $|\boldsymbol{\xi}|\pi\sigma^2$ of the oriented leg that issues from I, and $|I - J|$ is the distance between $\boldsymbol{\Gamma}_I$ and $\boldsymbol{\Gamma}_J$ (say between their centers). Since no two sites can be visited by the vortex more than once, no two segments can coincide and $|I - J| \neq 0$. Such a vortex is "self-avoiding". N is the number of legs, and $N\mu'$ is defined as before (Figure 5.5; the figure is drawn in two rather than three dimensions for easier inspection).

The vortex on the lattice of bond length h should be viewed as a rough description of a vortex with a fractal centerline; the description will be refined when h is decreased. Take a $M \times M \times M$ sublattice, and refine it by a factor 2, so that $h \to h/2$ and $M \to 2M$ (Figure 5.6). If the new, finer lattice has no new structure, then the vorticity spectrum is not self-similar and has a cut-off at $k \sim 1/L$, where L is a length scale typical of scales beyond which refinement introduces no new detail. If we do have

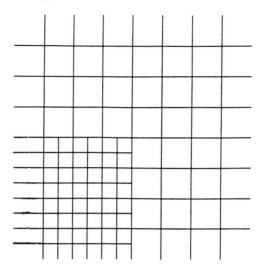

FIGURE 5.6. Refinement of the lattice.

a fractal vortex structure, the refinement of the lattice introduces new structure. Consider (1/8)th of the finer lattice; the resulting structure on this (1/8)th should be similar to the structure on the whole cruder lattice. The interaction energy $E_1 = \sum_I \sum_{J \neq I} \mathbf{\Gamma}_I \cdot \mathbf{\Gamma}_J / |I - J|$ will be 1/2 of the interaction energy on the whole crude lattice, as a result of the usual scaling relation. The number of bonds occupied by the vortex in the refined (1/8)th is larger than 1/2·of the number of bonds occupied by the vortex in the unrefined whole and the self energy of each bond has grown; thus its self-energy $(\mu' N)_{\text{new}} > \frac{1}{2}(\mu' N)_{\text{old}}$. The self-energy increases faster than the interaction energy. If one refines the lattice repeatedly, the self-energy will increase without bound, and there is no way to balance it out by bending the vortex on the lattice so as to decrease its interaction energy E_1. In this argument, changes in the shape of vortex cores and in how they fit into the lattice have not been taken into account.

The growth of the self-energy in the process of refining the description of a fractal vortex can be halted.[13] Suppose that after each refinement, each of the (1/8)ths of the lattice of linear dimension L is squeezed into a box of size $d^3 L^3 = \beta L^3$, $\beta < 1$ (β here is not $1/T$; the construction is partially patterned after the "β model"[14]). The squeezing is not a violation of incompressibility; it only implies that non-vortical fluid or non-stretching vorticity are expelled from among the stretching vorticity into the $(1 - \beta)L^3$ remaining volume and used to connect the stretching subvolumes to each other. Some vorticity is thus left behind and will no longer be stretched.

---

[13] A. Chorin, 1986.
[14] U. Frisch et al., 1978.

This is comparable with the decrease in u_n^2 in the Kraichnan derivation of the Kolmogorov law. It is immaterial here whether the stretching is viewed as a dynamical process occurring in time, as in the Kraichnan/Kolmogorov picture, or as a progressive revelation of finer structure in a fractal filament examined by an ever finer microscope.

In the squeezing, the energy $E = E_1 + \mu'N$ becomes $d \cdot E_1 + d^z \mu'N$ where $1 < z < 5/2$. Indeed, the length of each segment is multiplied by d and the distance between each pair of segments is multiplied by d, thus $E_1 \to d \cdot E_1$. Each entry in $\mu'N = \sum \sigma_I \tilde{E}(h/\sigma_I)$ is transformed in the form $\sigma_I \tilde{E}(h/\sigma_I) \to (\sigma_I/\sqrt{d})\tilde{E}\left((dh)/(\sigma_I/\sqrt{d})\right) = (\sigma_I/\sqrt{d})\tilde{E}(d^{3/2}h/\sigma_I)$. For h/σ_I small, $\tilde{E}(d^{3/2}h/\sigma_I) \sim d^3\tilde{E}(h/\sigma_I)$; for h/σ_I large, $\tilde{E}(d^{3/2}h/\sigma_I) \sim d^{3/2} \log(1/d)\tilde{E}(h/\sigma_I)$, and the claim is proved.

Thus, by picking an appropriate $d < 1$, one can decrease the growth of the self-energy relative to the interaction energy and allow the stretching to proceed. One can write $\beta = d^3 = 2^{D'-3}$, where D' is the fractal dimension of the set that surrounds the stretching vortex and that will ultimately contain almost all of the vorticity. Indeed, if $d = 1$, $\beta = 1$ and $D' = 3$; if at each refinement one keeps only a fraction β of the $(1/8)$th of box rather than all of $(1/8)$th $= (2^{-3})$th of it, the ratio of the volume one keeps to the available volume is $2^{D'-3}$.

An attempt has been made to determine D' numerically.[15] The ratio of the energy in a box on one scale to the energy in a box on another scale is determined by the Kolmogorov spectrum and will be estimated in Chapter 7; it requires an elaborate analysis. However, at each level of refinement one halves all the scales and one makes a step to the right in the graph of $E(k)$ as a function of k, i.e., $k \to 2k$. It is immaterial whether one views this step as part of a cascade or as a revelation of finer detail by a finer microscope. The ratio $Z(k)/E(k) = k^2$ is multiplied by 4. At each step of refinement one can estimate $\int \xi^2 dx$, $\int u^2 dx$ in the box, both of which depend on D' and on h/σ, set the ratio of the new ratio to the old ratio of these quantities equal to 4, make an assumption about h/σ and solve for D'. The problem is that one does not know how to chop off a finite piece of the lattice without affecting the outcome. The surface of the box is $O(L^2)$, where L is its linear scale, while the interaction between vortices dies out only as L^{-1}, thus no truncation at a finite distance makes real sense. Calculations with various ad-hoc truncations yield D' between 2.33 and 3. Most people seem to believe D is not far from 3, and thus D' is not far from 3. Note that even if $D = 3$, the measure of supp ξ in dimension 3 may be 0 and thus supp ξ may still be very sparse.

If the vorticity stretches into small volumes, one presumably obtains

---

[15] A. Chorin, 1986.

on each scale filament-like regions made up of folds upon folds of vortex filaments on finer scales. This is not unlike the situation in two dimensions, where the equilibria are made up of thin filaments on small scales. The difference is that the three-dimensional structures are themselves being folded and stretched. Each filament on a given scale is made up of finer filaments and participates in the stretching on cruder scales. One has vortices within vortices, down to scales when viscosity is important, as is consistent with the self-similarity of the Kolmogorov spectrum.

5.6. Vortex Cross-Sections

We have only considered so far what happens to the centerline of vortex filaments as the filaments stretch and fold. What happens to the cross-sections?

It is often observed in numerical calculations that stretching is more likely to occur in places when a lot of stretching has already occurred. This is compatible with the construction in the preceding section, where the stretching concentrates in ever smaller regions, and is thus more likely to occur in the neighborhood of past stretching. More generally, vorticity stretches vorticity, and there is more vorticity when stretching has already been extensive. The details of the stretching depend on the detailed configuration near the point on the filament that is being considered. One can model these remarks by the equation

(5.4)
$$\frac{d|\boldsymbol{\xi}|}{dt} = b(t)|\boldsymbol{\xi}|,$$

where $b = b(t, \omega)$ is a random coefficient. No claim is made that this equation is exact; for example, one cannot exclude by the argument above the possibility that $|\boldsymbol{\xi}|$ on the right-hand side should be replaced by $|\boldsymbol{\xi}|^{\alpha}$, $\alpha > 0$.

Equation (5.4) can be integrated:

$$\log|\boldsymbol{\xi}(t)| - \log|\boldsymbol{\xi}(0)| = \int_0^t b(t, \omega)dt.$$

It is plausible that $b(t_1, \omega)$, $b(t_2, \omega)$, $t_1 \neq t_2$, are independent. By the central limit theorem, the right side should then be a Gaussian random variable; if $|\boldsymbol{\xi}(0)|$ is a constant, one deduces that $|\boldsymbol{\xi}(t)|$ is a lognormal variable, i.e., that $\log|\boldsymbol{\xi}|$ is a Gaussian variable. This conclusion is reasonably well supported by numerical experiment.[16] Lognormal variables are quite wild, and their range is large. Since $|\boldsymbol{\xi}|$ is inversely proportional to the cross-section of a filament, that cross-section also has a lognormal distribution, and thus

---

[16] A. Chorin, 1982.

varies wildly. One way to check this conclusion is to observe that if one highlights, in a numerical calculation, those portions of the volume where $|\boldsymbol{\xi}|$ is large, one obtains a collection of unconnected tubular pieces. The filament portions in between the highlighted portions are wide and have a relatively small $|\boldsymbol{\xi}|$.[17]

The lognormality of $|\boldsymbol{\xi}|$ is compatible with the conclusions in the preceding section. If, at each level of refinement, the vortices are stretched by a factor λ, and the portions of the filaments outside the volumes $\beta 2^{-3}$ stop stretching, one obtains 8 pieces of vortex, each of length $(1-d)$, where $|\boldsymbol{\xi}| = |\boldsymbol{\xi}_0|$, ($|\boldsymbol{\xi}_0| =$ an initial value for $|\boldsymbol{\xi}|$), 8^2 pieces of length $\frac{1}{8}(1-d)$ where $|\boldsymbol{\xi}| = \lambda|\boldsymbol{\xi}_0|$, 8^n pieces of length $8^{-(n-1)}(1-d)$ where $|\boldsymbol{\xi}| = \lambda^n|\boldsymbol{\xi}_0|$, i.e., approximately equal lengths where $|\boldsymbol{\xi}|$ takes the value $\lambda^n|\boldsymbol{\xi}|$ and $\log|\boldsymbol{\xi}|$ takes the value $n\log\lambda|\boldsymbol{\xi}_0|$. Hence $\log|\boldsymbol{\xi}|$ varies moderately. Given the vagueness of the derivation of equation (5.4), this should count as good agreement.

The extreme variability of the vortex cross-sections opens the door to the possibility that the average vortex cross-section is non-trivial. If a vortex centerline can be defined and has dimension D_c, and if one looks at the cubes where stretching is still going on, with vortex radii satisfying $\sigma \leq h/2$ so as to fit on the lattice, one can readily see that in those cubes the dimension of the vorticity support dim supp $\boldsymbol{\xi} \to D_c$. However, in cross-sections where stretching has stopped, dim supp $\boldsymbol{\xi}$ may be larger than D_c. More complicated scenarios can be imagined. Inside the volumes where stretching is still occurring, the fraction of occupied bonds is $\beta_c = 2^{D_c-3} < 1$; one can imagine that as the vortex centerlines stretch, the cross-sections spread onto neighboring unoccupied bonds creating fractal cross-sections with eventually unbounded eccentricity. This process is often called "sheetification". Such cross-sections could then organize themselves into coherent objects, like the fingers in two dimensions do, in a local "inverse cascade" where the refinement may reverse itself. In Section 6.3, it will be shown that sheetification is necessary for energy conservation.

One can also conclude that the essential support of the vorticity may be much smaller and have smaller dimension than the original support of the vortex filament; $|\boldsymbol{\xi}|$ may converge to infinity on portions of a filament.

In summary, it is possible that portions of a vortex filament have a fractal dimension $D > D_c$, with "fat" portions contributing more than "thin" portions, and sections with unbounded eccentricity contributing more than sections where the dimension of the filament is converging to D_c. In general, the intersection of an object of dimension D with a plane has dimension $D - 1$;[18] the centerline intersects a plane on a set of dimension $D_c - 1$, which contains an infinite number of points if $D_c > 1$. The intersection of

---

[17]See, e.g., Z.S. She, 1991; P. Bernard et al., 1993.

[18]See, e.g., C.A. Rogers, *Hausdorff Measures*, Cambridge, 1970.

a vortex of dimension $D > D_c$ with a plane consists of an infinite collection of objects of various shapes and sizes. One can define the dimension of the cross-section of a vortex as $D - D_c$; that cross-section is usually a complicated fractal object.

In addition, we have seen that Brownian motion describes in space a fractal object that cannot be confined in a sequence of shrinking boxes as a connected curve; the interpolation formula of Section 2.5 shows that Brownian paths have large excursions, and their intersection with a small box generally consists of a collection of unconnected pieces. We shall see below that vortex centerlines have a relation to Brownian paths.

One may conclude that the geometry of vortex filaments is, not unexpectedly, extremely complicated; the words "vortex filament" should be in general interpreted in a highly generalized sense. If one views vortex filaments as "coherent structures", their "coherence" in three dimensions, unlike that of vortex structures in two dimensions, is incomplete. Of course, the coherent vortices generated by most numerical experiments smooth out the small-scale structure, and thus look more coherent than is possible in a self-similar flow with a spectrum that extends to $k = \infty$. A good numerical representation of what happens to a vortex tube has been given recently (Figure 5.7).[19]

5.7. Enstrophy and Equilibrium

It is quite obvious that the stretching described in the preceding sections has as a result a blow-up of the enstrophy $\langle \xi^2 \rangle$ (it is immaterial here whether this blow-up occurs in a finite or infinite time). We have made an analogy in Section 5.1 between entropy and enstrophy. It is therefore reasonable to ask whether there exists a maximum enstrophy state (be it a state of infinite enstrophy) that corresponds to a statistical equilibrium. The fact that the enstrophy is infinite should present no problem. By assuming that a small viscosity smoothes vortices on the Kolmogorov scale η one removes the smallest scales and renders the enstrophy finite. In many of the usual statistical mechanical arguments the dispersive smoothing due to quantum mechanics is used to make the infinite finite. For example, when it was said in Section 4.1 that Λ, the area of the sphere $H = E$, was proportional to the number of ways of arranging particles so as to achieve a certain macroscopic effect, it was implicitly assumed that points on $H = E$ within a small distance of each other (smallness being defined by quantum effects) are indistinguishable, or else the "number of ways" is infinite.

A heuristic way of seeing that quantization and viscosity should have qualitatively similar effects in statistical constructions can be seen from the

---

[19] J. Bell and D. Marcus, 1992.

following analogy. One standard way to construct a quantum version of a problem in classical mechanics runs through the "action principle". Given a problem with kinetic energy K and potential energy V, Hamiltonian $H = K + V$ and Lagrangian $L = K - V$, construct the action $A = \int_{t_1}^{t_2} L dt$, $L = L(q(t), p(t))$, t = time. A is a functional of the path between t_1 and t_2 that is stationary at the classical path, i.e., when $q(t), p(t)$ evolve as in the solution of the classical problem involving H. Indeed, the condition that A be stationary determines the classical evolution. To quantize this problem, one assigns to each path between the data at t_1 and the outcome at t_2 a weight proportional to $e^{(i/\hbar)A}$, where $\hbar$ is Planck's constant.[20] All paths contribute to the transmission of quantum amplitudes proportionally to their weight. The form of K creates weights of the form $e^{-ix^2/t\hbar}$, i.e., "Gaussians" with imaginary time. The effect is to smear the classical path. There is an obvious if imprecise analogy between this smearing and the smearing introduced when vortex trajectories are randomized to approximate diffusion.[21]

If indeed the end result of the stretching is a statistical equilibrium, one should be able to characterize it by an appropriate variational principle. In two dimensions, we have seen that the existence of an equilibrium is equivalent to the statement that the entropy $S = -\int_{D^N} f \log f \, d\mathbf{x}_1 \cdots d\mathbf{x}_n$ (or $S = -\int_D f_1 \log f_1 \, d\mathbf{x}$, where f_1 is the one particle density function, in the independent throws approximation) is maximum among all vorticity fields for which the energy $E = -1/4\pi \iint \xi(\mathbf{x})\xi(\mathbf{x}') \log |\mathbf{x} - \mathbf{x}'| d\mathbf{x} d\mathbf{x}'$ is constant. Similarly, one would expect that the maximization of $\int \xi^2 d\mathbf{x}$ among all $\boldsymbol{\xi}$ such that div $\boldsymbol{\xi} = 0$,

$$E = \frac{1}{8\pi} \int \int d\mathbf{x} d\mathbf{x}' \frac{\boldsymbol{\xi}(\mathbf{x}) \cdot \boldsymbol{\xi}(\mathbf{x}')}{|\mathbf{x} - \mathbf{x}'|} = constant,$$

would produce a sequence of $\boldsymbol{\xi}$'s with ever longer, more folded and more concentrated supports. Indeed, the calculation mentioned at the end of Section 5.5 is such a maximization, with piecewise tubular supp $\boldsymbol{\xi}$ (and possibly incorrect boundary conditions). These $\boldsymbol{\xi}$ could not be expected to be stationary solutions of the Euler equations, and the resulting equilibrium would be a statistical equilibrium.

The existence of an equilibrium would explain the reversibility of the flows once the Kolmogorov spectrum has been established (Section 3.3). The formation of the spectrum would correspond to an irreversible relaxation to equilibrium, the subsequent reversibility to the invariance properties of the equilibrium. The "universal equilibrium" of Section 3.1 would simply be a statistical equilibrium for vortex filaments.

---

[20] See, e.g., C. Itzykson and J. M. Drouffe, *Statistical Field Theory*, Cambridge, 1989.
[21] A. Chorin, 1991a.

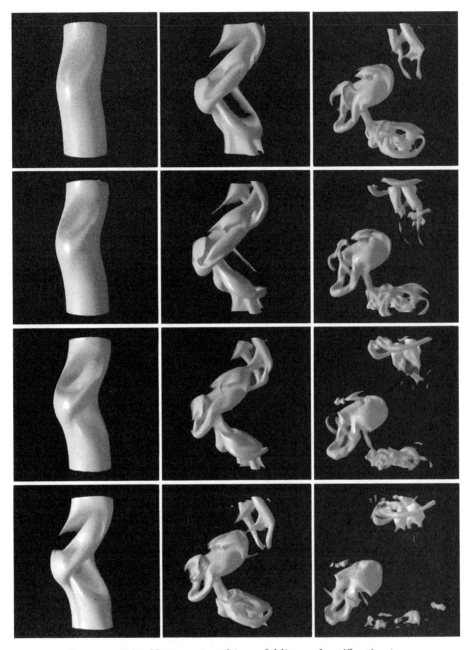

FIGURE 5.7. Vortex stretching, folding, sheetification.
[Reprinted with permission from J. Bell and D. Marcus,
Comm. Math. Phys. **147**, 371–394 (1992).]

However, even if the existence of such an equilibrium can be asserted, many additional questions must be answered. Is it unique? Can one conceive of a stretching and folding mechanism that asymptotically comes to a halt? How can a vortex equilibrium be reconciled with the increased dissipation that is characteristic of turbulence, and which can only be due to energy transport across the inertial range? What kind of measure is invariant (in other words, what ensemble is appropriate) in a situation where the lengths and shapes of the objects one is looking at are not invariant? We shall examine these questions in the next two chapters.

6
Polymers, Percolation, Renormalization

This chapter contains an assortment of facts and tools needed in the analysis of vortex equilibria in three space dimensions.

6.1. Spins, Critical Points and Metropolis Flow

One of the simplest statistical mechanics models is the Ising model in two space dimensions. It provides a good setting for presenting some useful ideas.[1]

Consider an $N \times N$ lattice in two space dimensions. On each lattice site (i,j), $1 \le i \le N$, $1 \le j \le N$, there is a "spin", i.e., a variable $\ell = \ell_{i,j}$ that can take on the values $\ell = +1$ ("spin up") or $\ell = -1$ ("spin down"). This system has 2^{N^2} "configurations" $C_1, \ldots$. The energy of a configuration C is

$$E(C) = -\sum_{i=1}^{N-1}\sum_{j=1}^{N} \ell_{i,j}\ell_{i+1,j} - \sum_{i=1}^{N}\sum_{j=1}^{N-1} \ell_{i,j}\ell_{i,j+1} \; ;$$

if $I = (i,j)$ is a multi-index, $E(C)$ can be written as $E(C) = -\sum_{I,J} \ell_I \ell_J$, where the summation is over neighboring sites only. The interaction between two neighboring sites with spins that point in the same direction contributes -1 to E, while the interaction between sites with spins that

---

[1]See, e.g., C. Thompson, *Equilibrium Statistical Mechanics*, 1988.

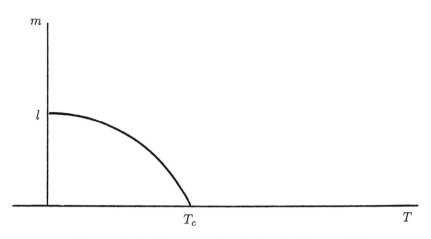

FIGURE 6.1. The magnetization in the Ising model.

point in opposite directions contributes $+1$. The system is assumed to be in equilibrium in a canonical ensemble, with $P(C) = e^{-\beta E(C)}/Z$, $Z = \sum_C e^{-\beta E(C)}$, $\beta = 1/T$; $\sum_C$ denotes a sum over all configurations.

The correlation function $\mathcal{R}(I, J)$ is

$$\begin{aligned}\mathcal{R}(I, J) &= \langle(\ell_I - \langle\ell_I\rangle)(\ell_J - \langle\ell_J\rangle)\rangle \\ &= \sum_C(\ell_I - \langle\ell_I\rangle)(\ell_J - \langle\ell_J\rangle)P(C) \,,\end{aligned}$$

where ℓ_I, ℓ_J are the spins at I, J in the configuration C. The definition of $\langle\ell_I\rangle$ is presumably obvious. The correlation length ζ is the order of magnitude of the distance $|I - J|$ required for $\mathcal{R}(I, J)$ to vanish. The magnetization m is $m = \lim_{N\to\infty}\langle\sum_I \ell_I\rangle/N^2 = \lim(\sum \ell_I P(C)/N^2)$. To simplify the presentation, we introduce a slight asymmetry into the problem by setting $\ell_I = +1$ at all the boundary points of the spin system.

For T small, $m \neq 0$; for T large, $m = 0$. We are particularly interested in the point T_c where m changes its behavior. The plot of m as a function of T looks as in Figure 6.1, with $T_c = 2.269\ldots(\beta_c = T_c^{-1} = 0.44\ldots)$. $m = 0$ for $T > T_c$; for $T < T_c$, $|T - T_c|$ small, $m \sim \tau^b$, $\tau = (1 - T/T_c)$, $b = 1/3$ (the usual notation is β instead of b, leading to confusion with $\beta = 1/T$). Near T_c, ζ blows up, with $\zeta \sim \tau^{-\alpha}$, $\alpha = 1$. The change in the behavior of m at $T = T_c$ is a "second-order phase transition" or "critical transition." $T = T_c$ is a "critical point", the exponents b, α are examples of "critical exponents". m is an "order parameter"; on one side of T_c there is order characterized by $m \neq 0$.

The blow-up of ζ and the existence of critical exponents near $T = T_c$ are closely related. If ζ is finite, $m = \langle\ell_I\rangle$ can be calculated by a finite sum of the form $\sum \ell_I P(C)$, $P(C) = e^{-\beta E(C)}/Z$, $Z = \sum_C e^{-\beta E(C)}$, where

C depends only on a finite number of spins; spins very far away do not matter. Under these conditions $\langle m \rangle$ is an analytic function of β and the non-analytic behavior at β_c cannot happen. If $\zeta = \infty$, m is described by an infinite sum, and infinite sums of analytic functions can behave in various odd ways. It follows in particular that the non-analytic behavior of m near T_c can be observed in full detail only when $N = \infty$.

For T near T_c, the system can have large fluctuations, i.e., there is a non-trivial probability attached to configurations C in which quantities such as E, the energy, differ non-trivially from their mean $\langle E \rangle$: if ζ is large, the number of effectively independent variables in the system is small and, as can be seen from Tschebysheff's theorem, the departure of their sum from the mean may be less than well bounded. In addition, if one divides the plane into regions where $\ell > 0$ and their complements where $\ell < 0$, one sees near T_c islands within islands of positive and negative spins, on all scales down to the scale of the lattice, distributed in a self-similar way when the scale of observation is varied. One hopes the reader is reminded of the properties of turbulence: one may expect turbulence to live near a critical point of some system.

Many systems have critical points as well as exponents that relate to various variables. Critical exponents are believed to have certain "universality" properties; if they refer to a lattice system they are generally independent of the specific lattice structure, and indeed of many of the details of the system other than the dimension of the space in which it is imbedded and the dimension of the "order parameter" which marks the transition.

Consider now the problem of constructing a time-dependent system for which the Gibbs distribution of spin configurations for N finite is an equilibrium that is approached as $t \to \infty$. This is the converse of the problem we have studied in earlier sections, in which we were given a time-dependent system and were looking for its statistical equilibria. A (non-unique) recipe for constructing such a time-dependent system is given by "Metropolis flow": Given a configuration C_n, construct a new configuration $C_{n+1} = M_n C_n$, where M_n is a (usually random) operator; then construct $C_{n+2} = M_{n+1} C_{n+1}$, etc. The result is a walk on the space of configurations; that space is related to the space in which the spins live like phase space of Chapter 4 is related to the space in which the vortices live. The sequence $C_1, C_2, \ldots, C_n, \ldots$ constructed in this way is a Markov chain if the probability of hitting a configuration C_{n+1} given C_n is independent of all the configurations prior to C_n, i.e., if the walk has no memory. Suppose we have a Markov chain of states that satisfy the following conditions:

(1) The chain is "ergodic": Given any pair C_-, C_+ of configurations that the system can be in (for example, in the spin case, any two

of the 2^{N^2} configurations that can be constructed), there is a non-zero probability that one can reach C_+ in a finite number of steps starting from C_-.

(2) The chain satisfies the "detailed balance condition": If $P(i \to j)$ is the probability that in one step one goes from configuration C_i to configuration C_j, and if $P(C_i)$ is the probability of C_i, then

$$\frac{P(i \to j)}{P(j \to i)} = \frac{P(C_j)}{P(C_i)} \ .$$

In particular, one is more likely to go from a less probable to a more probable state than the opposite. It is important to note that in the case of the Gibbs probability $P(C_j)$ the hard-to-calculate partition function Z has cancelled out of this condition.

If one views the sequence $C_n, C_{n+1}, \dots$ as a time evolution, then that time evolution leaves invariant the canonical measure $e^{-\beta E}/Z$. As the number of steps tends to infinity, each configuration is visited with a frequency that approximates its probability; the sequence of steps can be used to calculate averages with respect to $P : n^{-1} \sum_{i=1}^{n} f(C_i) \to \langle f(C) \rangle$. Near T_c, one can expect the values of $f(C_i)$ to have substantial variance, and thus the accuracy of such a calculation to be poor.

As an example of an ergodic sequence of transformation that satisfies the detailed balance condition, consider the Ising model; pick a site (i, j) at random; calculate the change in energy ΔE that would result from the flip $\ell_{i,j} \to -\ell_{i,j}$. Carry out the flip with a probability $\tilde{p} = \min(1, e^{-\beta \Delta E})$ and keep $\ell_{i,j}$ at its previous value with probabilaity $1 - \tilde{p}$. Ergodicity is obvious. The detailed balance condition follows from

$$\frac{\min(1, e^{-\beta \Delta E})}{\min(1, e^{\beta \Delta E})} = e^{-\beta \Delta E} = \frac{e^{-\beta(E + \Delta E)}}{e^{-\beta E}} = \frac{e^{-\beta(E \text{with flip})}}{e^{-\beta(E \text{without flip})}} \ .$$

6.2. Polymers and the Flory Exponent

Consider again a cubic lattice in a d-dimensional space, and a connected sequence of N bonds, with no site separating more than two bonds in the sequence (Figure 6.2). Such a sequence is a "self-avoiding walk", or SAW for short. (The lattice vortices constructed in Section 5.3 were self-avoiding.) Assume that all SAW's have equal probabilities. If one views two SAW's that can be obtained from each other by rigid rotation or translation as being identical, there is, on a square lattice in two space dimensions, one possible one-step SAW, three two-step SAW's, $9 = 3^2$ three-step SAW's, but fewer than $3^3 = 27$ four-step SAW's, since visits to earlier sites are forbidden.

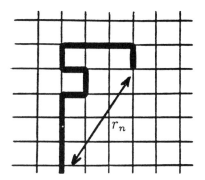

FIGURE 6.2. A self-avoiding walk.

A collection of equal-probability N-step SAW's is known as a "polymer", because of its uses in theoretical chemistry. Let r_N be the end-to-end straight-line distance between the beginning and the end of a SAW; we claim that

$$(6.1) \qquad \bar{r}_N \equiv \langle r_N^2 \rangle^{\frac{1}{2}} \sim N^{\mu} \quad \text{for large } N ,$$

where μ is the Flory exponent. (Note that we're using the letter μ both for chemical potential and for Flory exponent. This usage is common, and the meaning of μ should be apparent from the context.) The calculation of μ is related to problems in spin statistics,[2] but we shall not exploit this relation here. It should be clear that $\mu \geq \frac{1}{2}$; $\mu = \frac{1}{2}$ is the Brownian walk value (Section 3.4), and the constraint of self-avoidance should decrease the number of folds, straighten the walk and increase μ. μ is a critical exponent, and is properly defined only in the limit $N \to \infty$,

$$\mu = \lim_{N \to \infty} \frac{\log \bar{r}_N}{\log N}$$

Flory gave an argument[3] from which one can deduce $\mu = 3/(d + 2)$, where d is the dimension of the space. Flory's argument makes, as far as one can see, no sense whatsoever; it is a striking case of right answers derived by wrong arguments. For $d = 1$, the only SAW is the line itself and $\mu = 1$. For $d = 2$ Flory's value is $\mu = \frac{3}{4}$, and is well verified numerically. For $d = 3$, Flory's value is $\mu = \frac{3}{5} = 0.6$; the best numerical estimate is $\mu = 0.588$.[4] For $d = 4$, $\mu = 0.5$, which is exactly right; for $d > 4$, μ remains at 0.5, contrary to Flory's formula.

The value $\mu = 0.5$ when $d \geq 4$ is significant: $\mu = 0.5$ is the exponent for Brownian motion, which is not self-avoiding. The fractal dimension

---

[2] See, e.g., S. K. Ma, *Modern Theory of Critical Phenomena*, Benjamin, 1976.

[3] See, e.g., P. G. de Gennes, *Scaling Laws in Polymer Physics*, Cornell, 1971.

[4] A. Sokal, 1991, private communication.

of Brownian motion is 2. It is unlikely that in a space of dimension $d \geq 4$ two Brownian motion paths will intersect, and it is therefore unlikely that a single Brownian motion will intersect itself.[5] Thus when $d \geq 4$, there should be little difference between an equal-probability SAW and a Brownian motion, and $\mu = 0.5$. It is often the case that geometrical constraints become inoperative in a space of large enough dimension; the smallest dimension d that is large enough in this sense is the "upper critical dimension" for the problem. Thus 4 is the upper critical dimension for the polymer problem.

Equation (6.1) can be inverted:

(6.2)

$$N \sim \bar{r}_N^{1/\mu}$$

Thus $D = 1/\mu$ is the fractal dimension of a polymer. When $d = 3$, $D = 1/\mu \cong 1.70$; more crudely, $D \sim 1.66 = 5/3$ if $\mu = 3/5$, the Flory value.

Equation (6.2) is sufficient to find the form of the density correlation function for a polymer.[6] Suppose h, the lattice spacing, is small and suppose we average the number of bonds that belong to the polymer (= "monomers") in a region large compared to h^d, obtaining a monomer density ρ. Suppose we consider a dilute, homogeneous suspension of polymers, and calculate $\langle \rho(\mathbf{x})\rho(\mathbf{x}+\mathbf{r}) \rangle$ for $|\mathbf{r}| = r >> h$ but small compared with N^μ. Pick $\mathbf{x}$ on the polymer (otherwise the contribution of $\mathbf{x}$ to $\langle \rho(\mathbf{x})\rho(\mathbf{x}+\mathbf{r}) \rangle$ is zero), and consider a sphere S_r of radius r around $\mathbf{x}$. There are $\sim r^{1/\mu}$ monomers in S_r; there are $\sim r^{(1/\mu)-1}$ monomers between r and $r + dr$; their density between r and $r + dr$ is $\sim r^{(1/\mu)-1}/r^{d-1} = r^{(1/\mu)-d}$. Thus

$$\langle \rho(\mathbf{x})\rho(\mathbf{x}+\mathbf{r}) \rangle \sim r^{(1/\mu)-d} , \quad h << r << N^{1/\mu} .$$

The divergence of the right-hand side at $r = 0$ is not a problem: the formula is not expected to hold at $r = 0$. The formula applies to ever smaller values of r as $h \to 0$.

Let $\mathcal{R}(\mathbf{r}) = \langle \rho(\mathbf{x})\rho(\mathbf{x}+\mathbf{r}) \rangle$. The Fourier transform $\psi(\mathbf{k})$ of $\mathcal{R}(\mathbf{r})$ is $O(k^{-1/\mu})$ for large k; if $d = 3$, $D = 1/\mu \cong 5/3$, and $\psi(\mathbf{k}) = O(k^{-\frac{5}{3}})$.

Beware of claims that this calculation provides a derivation of Kolmogorov's law! ρ is a scalar, while $\boldsymbol{\xi}$ is a vector; we shall see below that the difference between vector and scalar correlations is large. In addition, the fractal dimension of the support of ρ is $\sim 5/3$, which is too low for the support of $\boldsymbol{\xi}$. Observe however that if one writes $\psi(\mathbf{k}) = O(k^{-\gamma})$, then $\gamma = D$, and $\frac{d\gamma}{dD} = 1$ is positive, as discussed in Chapter 3.

The exponent μ can be evaluated by constructing a Markov chain that spans the set of polymer configurations[7]: Consider the set of automor-

[5]M. Aizenman, 1982.

[6]See P. G. de Gennes, *loc. cit.*

[7]M. Lal, 1969; N. Madras and A. Sokal, 1988.

phisms of the cubic lattice, i.e., the set of transformations that map the lattice on itself. Each such automorphism can be represented by an orthogonal matrix with integer entries. An orthogonal matrix has orthonormal columns, and thus has only one non-zero entry in each column and on each line, the non-zero entry being $+1$ or -1. There are 48 such matrices in three-dimensional space.

Pick a SAW of N steps. Pick one of its ends and label it O, the "floating end". Pick a site P on the SAW at random, with equal probability for each site to be picked; rotate the piece of the SAW between O and P by one of the automorphisms, also picked at random with equal probabilities. If the new SAW is self-avoiding, accept it; if it is not self-avoiding, identify the new configuration with the old configuration. This "rotation" defines the new SAW in the Markov chain of SAW's. It is easy to check that the sequence is ergodic (any SAW has a chance to be folded into any other SAW) and satisfies the detailed balance condition for polymers (SAW's with equal probabilities). This Markov chain can then be used to calculate the Flory exponent.

Consider the sequence of folded SAW's as a time-dependent flow; the resulting Lal-Madras-Sokal flow has interesting properties. This flow consists of a sequence of foldings. One could think that a sequence of foldings will keep on increasing the fractal dimension of SAW's until the upper bound $D = 3$ is reached. This does not happen. At $D = 1.70 \sim 5/3$, the folding rearranges the N-step SAW's within a family whose dimension remains fixed. If the constraint of self-avoidance is removed, and each folded connected set of bonds is accepted in the folding, the end result is a family of Brownian walks with $D = 2$. By analogy, these remarks make more plausible the idea that hydrodynamic stretching and folding can generate an equilibrium ensemble.

Note finally that in a sufficiently dense assembly of polymers the exponent μ changes value. We shall never have to worry about this situation in a hydrodynamical problem where intermittency reigns.

6.3. The Vector-Vector Correlation Exponent for Polymers

Polymers have a second critical exponent,[8] directly relevant to vortex statistics, and defined as follows: orient the polymer, so that all its legs are vectors. Pick a lattice site on the polymer labeled I, and let $\mathbf{\Gamma}_I$ be the vector along the polymer issuing from I (as in Section 5.4). Consider the

---

[8]A. Chorin and J. Akao, 1991.

sum

$$S_r = \left\langle \sum_{|I-J| \leq r} \mathbf{\Gamma}_I \cdot \mathbf{\Gamma}_J \right\rangle ,$$

i.e., consider all the vectors $\mathbf{\Gamma}_J$ on the polymer within a distance r of I, add $+1$ if $\mathbf{\Gamma}_I$ and $\mathbf{\Gamma}_J$ are parallel, -1 if they are anti-parallel, 0 otherwise. S_r satisfies the relation

(6.3)
$$S_r \sim r^{1/\tilde{\mu}} = r^{\tilde{D}}$$

for r large enough. In a dilute suspension of polymers one must be careful to make r small enough so that there is no other polymer within r of I; r must also not be so large that the polymer has a substantial probability of ending within a distance r from I. $\tilde{D}$ is defined by $\tilde{D} = 1/\tilde{\mu}$, and is not a fractal dimension. $\tilde{D}$ is the vector analogue of $D = 1/\mu$ of the scalar case.

If one identifies the vector along the polymer with a vorticity vector, then knowledge of $\tilde{D}$ allows one to calculate the resulting vorticity correlation and spectrum, and then the energy spectrum. Indeed, following the steps in the calculation of the density correlation function in the previous section, we find that there are $\sim r^{\tilde{D}-1}$ positive contributions to S_r between r and $r + dr$; their density is proportional to $r^{\tilde{D}-1}/r^2 = r^{\tilde{D}-3}$; the correlation function $\langle \mathbf{\Gamma}(\mathbf{x}) \cdot \mathbf{\Gamma}(\mathbf{x} + \mathbf{r}) \rangle \sim \langle \boldsymbol{\xi}(\mathbf{x}) \cdot \boldsymbol{\xi}(\mathbf{x} + \mathbf{r}) \rangle \equiv Q(\mathbf{r})$ is $Q(\mathbf{r}) \sim r^{\tilde{D}-3}$ for $h \ll r \ll N^\mu$. (Note that μ, not $\tilde{\mu}$, characterizes the length of the polymer.) The Fourier transform of $Q(\mathbf{r})$ is $O(k^{-\tilde{D}})$. The vorticity spectrum $Z(k)$ is obtained by integrating that Fourier tranfsorm over a sphere of radius $k = |\mathbf{k}|$, yielding $Z(k) = O(k^{-\tilde{D}+2})$. The energy spectrum is $E(k) = \frac{Z(k)}{k^2} = O(k^{-\tilde{D}})$ for large k ($k \ll h^{-1}$, $h \to 0$, where h is the lattice spacing).

Clearly, $S_r \leq 2r$ (Figure 6.3); if the lattice spacing is h, the number of positive contributions to S_r is always $\leq 2r/h$, therefore $\tilde{D} \leq 1$. If $P(\mathbf{\Gamma}_I \cdot \mathbf{\Gamma}_J > 0) \geq P(\mathbf{\Gamma}_I \cdot \mathbf{\Gamma}_J < 0)$, i.e., if the "vortex legs" are more likely to be parallel than to be anti-parallel, then $\tilde{D} \geq 0$. A numerical calculation[9] has yielded $\tilde{D} = 0.37 \pm 0.02$. A series expansion carried out to first order[10] gives $\tilde{D} = 0.25 + 0(1)$. The corresponding spectrum $O(k^{-\gamma})$, $\gamma = \tilde{D}$, has γ smaller than the Kolmogorov value; the dimension of supp $\boldsymbol{\xi}$ is $\sim 5/3$, also too small for hydrodynamics; as pointed out in Chapter 3, these two facts are self-consistent.

Note that when supp $\boldsymbol{\xi}$ is topologically one-dimensional, i.e., smoothly deformable into a curve, γ and $D = $ dim supp $\boldsymbol{\xi}$ do not always grow together, unlike what happens in the scalar case and unlike the general trend outlined with caveats in Chapter 3. Indeed, compare $\tilde{D} = 2, \gamma = 0$ for

---

[9] A. Chorin and J. Akao, 1991.
[10] R. Zeitak, 1991.

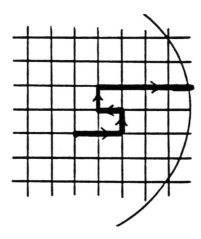

FIGURE 6.3. Estimation of $\tilde{D}$.

Brownian walks, $D \equiv 5/3, \gamma \sim 0.37$ for polymers. The inequality $\tilde{D} \leq 1$ holds for any supp $\boldsymbol{\xi}$ that is topologically one-dimensional, as can be seen from Figure 6.3. Since $\langle \mathbf{u}^2 \rangle = \int E(k)dk$, $\langle \mathbf{u}^2 \rangle$ is not finite. A comparison with the discussion of vector capacity in Section 5.4 suggests that a set which is topologically one-dimensional never has a finite vector capacity. To make the comparison, one has to assert that a statistically translation invariant random set, $S = S(\omega)$, $\omega \in$ probability space, which for every ω can support a finite vorticity as defined in Section 5.4 so that $\int \mathbf{u}^2 d\mathbf{x} < +\infty$, can also support for each ω a vorticity field $\boldsymbol{\xi} = \boldsymbol{\xi}(\mathbf{x}, \omega)$ that is statistically translation invariant and has $\langle \mathbf{u}^2 \rangle$ finite. This assertion is plausible, and does not contradict the claim in Section 5.4: it was claimed there that given $\varepsilon > 0$, one can find a set of dimension $1 + \varepsilon$ that has positive vector capacity, not that every set of dimension $1 + \varepsilon$ has positive vector capacity. One conclusion from this argument about $\tilde{D}$ is that one cannot have any part of supp $\boldsymbol{\xi}$, the support of the vorticity, be topologically one-dimensional, and thus that in the stretching process as described in Section 5.6 "sheetification" is likely to occur; more generally, one must have $D > D_c$, where $D = \dim \operatorname{supp} \boldsymbol{\xi}$, $D_c = $ dimension of the centerline of the vortex filament.

6.4. Percolation

Percolation theory[11] is the study of the collective behavior of random objects, motivated originally by the question whether the array of random passageways that occur in a porous medium ever assembles into a macroscopic passageway. An example of a percolation problem is the following

---

[11]See, e.g., D. Stauffer, *Introduction to Percolation Theory*, Taylor & Francis, 1985.

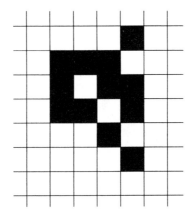

FIGURE 6.4. Percolation clusters.

"plaquette" percolation problem: Consider a lattice in the plane, and consider the squares ("plaquettes") defined by its bonds (Figure 6.4). Go from plaquette to plaquette, and in each one, cover the plaquette with black ink with probability p, or leave it white with probability $1 - p$, with what is done in one plaquette being independent of what had been done in the previous ones. Black plaquettes that have a common bond in their boundary are neighbors; black plaquettes that touch each other at one point are not neighbors. A collection of black plaquettes that can be connected to each other through black ones that are neighbors form a "percolation cluster".

The first question one wishes to answer is: what is the probability $\mathbf{P}$ that the plane contains an infinite cluster (one with an infinite number of black plaquettes)? If the black plaquettes are thought of as conducting and the white ones as insulating, the question can be rephrased as, What is the probability that the array is globally conducting? Clearly $\mathbf{P}$ is a nondecreasing function of p: the more black plaquettes there are the more likely it is that some fraction of them coheres into an infinite cluster. What may be unexpected is that $\mathbf{P}(p)$ looks as follows: $\mathbf{P} = 0$ for $p < p_c = 0.5927\ldots$, $\mathbf{P} = 1$ for $p > p_c$; for $p = p_c$ one can find percolation clusters of arbitrarily large size, but not necessarily of infinite size. In the conductor/insulator interpretation, the material is insulating for $p < p_c$ and conducting for $p > p_c$. p_c is a "percolation threshold". The point $p = p_c$ is a critical point in the sense of Section 6.1. To have a sharp transition at $p = p_c$ the lattice has to be infinite; a finite lattice has a small but non-zero probability to be conducting even when $p < p_c$; this probability decreases with lattice size. The probability $\tilde{p}$ that a given plaquette belongs to an infinite cluster can be viewed as an order parameter: $\tilde{p} = 0$ for $p < p_c$, $\tilde{p}$ is finite for $p > p_c$. An appropriate correlation length diverges at $p = p_c$; the order parameter grows for $p > p_c$ like a power of $p - p_c$.

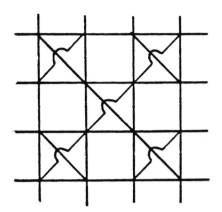

FIGURE 6.5. Lattice dual to a square lattice.

The boundary of a percolation cluster in the problem just discussed is the set of bonds that have black members of the cluster on one side and white elements on the other. Typically, the boundary consists of several unconnected components, because a percolation cluster typically has "holes". The component that can be connected to infinity through white neighboring or near-neighboring plaquettes is the cluster "hull". (The reason why near-neighbor plaquettes are allowed can be seen by drawing a picture.) The number of bonds N in a hull increases with the linear dimensions of the cluster[12] according to $N \sim L^{7/4}$. The fractal dimension of the hull is therefore $D = 7/4$. An appropriate change in the definition of the hull (that we shall not need) can result in $D = 4/3$, the same dimension as that of a polymer in two-dimensional space.

The plaquette percolation problem is identical to the site percolation problem, in which the sites or the lattice are either occupied (probability p) or empty [probability $(1 - p)$]. Two neighboring occupied sites are assumed to be connected, and the question is whether there exists an infinite connected cluster. Bond percolation is different: assume the bonds in the square lattice are either conducting (probability p) or not [probability $(1-p)$]. Is there an infinite conducting cluster? Yes, if $p > p_c$; no, if $p < p_c$, with $p_c = 1/2$ for the square lattice. A bond percolation problem can be reduced to a site percolation problem on the "dual" lattice, constructed as follows: mark the center of each bond with a cross; connect crosses by a line if the corresponding bonds touch. A bond on a square lattice touches six other bonds, and thus a site on the dual lattice is connected to six other sites. The lattice dual to a square lattice is drawn in Figure 6.5.[13]

On this dual lattice $p_c = 1/2$; note that by adding links we have made it

[12]H. Saleur and D. Duplantier, 1987.
[13]See, e.g., G. Grimmett, *Percolation*, 1979.

easier to have an infinite connected cluster and reduced p_c. The plaquette version of this dual site percolation problem is as follows: place a point of coordinates (i, j) in the middle of each plaquette [thus the lattice sites are at $(i \pm \frac{1}{2}, j \pm \frac{1}{2})$]. If $(i+j)$ is even, black plaquettes at (i, j), $(i+1, j+1)$, and at (i, j) and $(i - 1, j - 1)$ (to the northeast and southwest) are viewed as connected, while black plaquettes to the northwest and southeast are not. If $(i + j)$ is odd, the converse holds: (i, j) is connected to the northwest and southeast.

6.5. Polymers and Percolation

In this section we present an example of a relationship between percolation and polymer statistics,[14] which constitutes a simple version of a more elaborate structure that will be discussed in the context of vortex dynamics.

Consider SAW's in the plane, open (i.e., with dangling end points) or closed (i.e., forming closed loops that divide the plane into an interior and an exterior). Assume that the bonds in the SAW attract each other, with a short-range attraction that is operative across one lattice spacing only. The easiest way to incorporate this attraction into the statistics is to give each SAW a probability that is proportional to $e^{n/T}$, where n is the number of close encounters in the SAW; this probability replaces the equal probabilities of the preceding non-interacting polymer problem. The fractal dimension of the resulting polymers is independent of whether they are closed or open.

The effect of this attraction on the statistics (and in particular on the fractal dimension) of the polymer is as follows: there is a temperature T_c such that for $T > T_c$ the effect is negligible; for $T < T_c$ the effect is catastrophic—the polymer folds up tightly and its dimension increases to 2, the maximum possible for a planar lattice. We now show that at $T = T_c$ there is an intermediate state, whose statistics are those of the hull of a percolation cluster at the percolation threshold, and in particular $D = 7/4$.

We shall take for granted that only three states are possible: the unaffected state with $D = 4/3$, the folded state with $D = 2$ and the intermediate state. If we find a state with $4/3 < D < 2$ it must be the intermediate state. We shall examine the structure of that intermediate state on a hexagonal lattice, relying on "universality" to carry the conclusions over to other lattices.

One cannot in general construct a SAW of N steps on a two-dimensional lattice by simply walking at random and avoiding previously occupied sites,

---

[14] A. Weinrib and S. Trugman, 1985; A. Coniglio, N. Jan, I. Majid and H. E. Stanley, 1987.

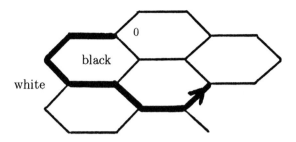

FIGURE 6.6. A smart walk on a hexagonal lattice.

because there is a high probability of being trapped in a loop from which there is no exit that does not cross a previously occupied location. Imagine therefore a smart walk in the plane that avoids not only previously occupied sites but also avoids traps. On a hexagonal lattice one can avoid traps through the use of local information.

Suppose a walker starts at the origin 0 (Figure 6.6) and wanders at random. At each site he has two choices, and picks one of them. If he is ever brought back to the origin, he accepts his fate and ends the walk, which then becomes closed. Each step is a boundary between two hexagons. If the boundaries of neither hexagons have yet been visited, the walker is free to do as he pleases; each of his choices has probability $1/2$. If one of the hexagons has a boundary bond that has already been visited, then there may be only one way of avoiding a trap and the resulting step has probability 1. An N-step walk has then probability $Z^{-1} \cdot 2^n$, where $Z^{-1} = 2^{-N}$ is a factor common to all walks, and n is the number of close encounters on a hexagon. However, $2^n = e^{(\log 2)n} = e^{n/T_c}$, where $T_c = 1/\log 2$ is the critical temperature.

In the first step, paint the hexagon on one side black and the other white. In each step that is unconstrained by the need for self-avoidance or for avoiding traps, assign to the hexagon in front of the last step the color black or white with a probability $1/2$ for each choice, independently of the previous assignments, in such a way that black remains on one side of the walk and white on the other. Whenever the choice of step is forced, the color or the hexagon in front has already been chosen. Thus, as the walk proceeds, it generates the outer layers of a percolation cluster of which it forms the hull. The resulting dimension is $D = 7/4$, if one believes in universality, and thus the walk generates the intermediate state of the polymer. That was the claim. The probability of a black hexagon is $1/2$, and this is the critical probability for the hexagonal lattice. Thus the intermediate polymeric state traces out cluster hulls at the percolation threshold.

6.6. Renormalization

At its simplest, renormalization is a technique for lumping together variables, thus reducing their number, in such a way that the statistical behavior of the system that those variables describe is not altered. The basic observation that is needed in the analysis below is that the partition function of a system in equilibrium, $Z = \sum_s e^{-\beta E_s}$, contains all the information needed to describe the system, so that a transformation that changes parameters in the system but leaves Z unchanged also keeps the physics unchanged. Indeed, we already know that $\langle E \rangle = -\frac{\partial}{\partial \beta} \log Z$. We also saw in Section 4.1 that the entropy S equals $\beta \langle E \rangle + \log Z$; $\beta \langle E \rangle + \log Z = \frac{\partial}{\partial T}(T \log Z)$. All the other thermodynamic quantities can be derived from Z.

To explain the idea of renormalization, consider the one-dimensional Ising model.[15] Spins $\ell_i = \pm 1, i = 1, \ldots, N$ sit on the sites of a one-dimensional lattice, with the energy E of a configuration C given by $E(C) = -\sum_1^{N-1} \ell_i \ell_{i+1}$ and the probability $P(C) = Z^{-1} e^{-E(C)/T}, Z = \sum_C e^{-E(C)/T}$.

The summation over C is a summation over the values of $\ell_1, \ell_2, \ell_3, \ldots$. Perform first the summation over every second spin, say, ℓ_2, ℓ_4, ℓ_6. Each one of these can take on the value ± 1, and interacts only with its neighbors to the right and left. After summation over $\ell_2, \ell_4, \ell_6, \ldots$, one obtains

$$Z = \sum_{\ell_1, \ell_3, \ell_5, \ldots} \left(e^{\beta(\ell_1 + \ell_3)} + e^{-\beta(\ell_1 + \ell_3)} \right) \cdot \left(e^{\beta(\ell_3 + \ell_5)} + e^{-\beta(\ell_3 + \ell_5)} \right) + \cdots .$$

Now try to recast Z as a partition function for a system in which the variables are $\ell_1, \ell_3, \ldots$ and the variables summed over have disappeared. Suppose one can find a function $F(\beta)$ and a new temperature $\tilde{\beta}$ such that

$$(6.4) \qquad e^{\beta(\ell_1 + \ell_3)} + e^{-\beta(\ell_1 + \ell_3)} = F(\beta) e^{-\tilde{\beta} \ell_1 \ell_3}$$

for all values ± 1 of ℓ_1 and ± 1 of ℓ_3. Then

$$Z(\beta, N) = \sum_{\ell_1, \ell_3} (F(\beta))^{N/2} e^{-\tilde{\beta}(\ell_1 \ell_3 + \ell_3 \ell_5 + \cdots)}$$

$$(6.5) \qquad\qquad = (F(\beta))^{N/2} Z(\tilde{\beta}, N/2) .$$

We have expressed the partition function for N spins at an inverse temperature β in terms of the partition function for $N/2$ spins at an inverse temperature $\tilde{\beta}$, provided equation (6.4) can be satisfied. To check whether

---

[15]See, e.g., D. Chandler, *loc. cit.*.

(6.4) can be satisfied, plug into it the values of ℓ_1, ℓ_3, obtaining

$$2\cosh 2\beta = F(\beta)e^{\tilde{\beta}}$$
$$2 = F(\beta)e^{-\tilde{\beta}} ,$$

and hence,

(6.6) $$\tilde{\beta} = \tfrac{1}{2}\log\cosh 2\beta ,$$

(6.7) $$F(\beta) = 2\sqrt{\cosh 2\beta}.$$

The expressions for $\langle E \rangle, S$, etc., involve $\log Z$, not Z standing alone. Let $\phi = N^{-1}\log Z$. The recursion relation (6.5) reads, in terms of ϕ:

(6.8) $$\phi(\tilde{\beta}, N/2) = 2\phi(\beta, N) - \log F(\beta)$$

Equations (6.6), (6.7), (6.8) are the "renormalization group equations" for the system. They express $\log Z$ for a system of $N/2$ particles at an inverse temperature $\tilde{\beta}$ in terms of $\log Z$ for N particles at β. The number of particles has been reduced, and the system has been changed, but the statistical properties of the original system have not been lost. Equation (6.6) expresses the "parameter flow", i.e., the way the parameter β evolves as the system is "renormalized", i.e., reduced in size. One can readily check that $\tilde{\beta} < \beta$; in each step the temperature increases; eventually one reaches a neighborhood of $T = \infty$ when all states are equally likely, $Z \cong \sum_{2^N \text{states}} 1 = 2^N$, $\phi = \log 2$. Marching back, one can in principle calculate ϕ for a finite T.

Unfortunately, this problem is uncharacteristically simple, and nothing interesting happens in it, because there are no critical points. In a two-dimensional Ising problem, in which there is a critical point, the analogues of equations (6.4) cannot be solved exactly, and one produces only an approximate renormalization group.

Each renormalization step halves the system, and thus halves the correlation length ζ because two spins have been replaced by one (see Section 6.1). At a critical point $\zeta = \infty$, and therefore $\beta_c = 1/T_c$ should be a fixed point of the renormalization group transformations. This is indeed the case if things are done right. A critical point is an unstable point of the renormalization "flow", because if $T \neq T_c$, ζ is finite, and successive halvings will eventually make it small. We shall present a renormalization transformation for a system with a critical point in the next section, where the system is a vortex system in the plane.

6.7. The Kosterlitz-Thouless Transition

Consider[16] a plane, with vortices of strengths $\pm\Gamma$, confined, as in Chapter 4, to a finite region $\mathcal{D}$ with area $|\mathcal{D}|$, in an equilibrium described by a canonical distribution. To begin with, Γ is assumed given, so that in particular there are no vortices whose strength has absolute value less than Γ. At very low temperatures only low-energy states can exist with significant probability. An example of a low-temperature state is a pair of vortices of opposing sign separated by a small distance r. If $r = 0$ in such a pair there is of course no vortex at all.

We suspend the conservation of circulation, and suppose that the system has a mechanism for creating vortex pairs ab nihilo; there will always be an equal number of positive and negative vortices. At low T one can have a few pairs with small separation. At slightly higher T there will be more pairs; the pairs can "screen" each other, i.e., arrange themselves so as to nearly neutralize each other, giving rise to the possibility that the separation r can increase. As the number of pairs increases with T, the separations r grow until at a $T = T_c$ there arise a number of "free" vortices, divorced from their partners in the pairs. The transition to a system with free vortices is the "Kosterlitz-Thouless transition". If instead of vortices one thinks of electric point charges, which have the same energy function as vortices, one sees that at $T = T_c$, the system changes from a neutral collection of pairs of charges, which is a dielectric (non-conducting) medium, to a system of isolated charges of both signs which, when subjected to an electric potential, will carry a current, and will thus be a conductor. This transition can also be used to model the transition from a superfluid to a normal fluid state in thin films of helium, as well as other interesting two-dimensional phenomena. As long as the vortices live in pairs, the separation into patches of opposing signs we saw in Chapter 4 cannot occur.

We assume the N vortex systems (N even) is described by a Hamiltonian

$$H = -\sum_i \sum_{j \neq i} \Gamma_i \Gamma_j \log|\mathbf{x}_i - \mathbf{x}_j| + N\mu \ ,$$

where the $\mathbf{x}_i$ are the locations of the vortices, $\Gamma_i = \pm 1$, the logarithm is suitably smoothed near the origin, and the usual factor $(4\pi)^{-1}$ and the prime on μ have been omitted as irrelevant. We shall have occasion to think about the sign of the log, and to avoid ambiguity we assume that $\mathcal{D}$ is large enough so that a distance $|\mathbf{x}_i - \mathbf{x}_j| = 1$ can be viewed as small and the log is positive. The easiest way to enforce this condition is to replace $\log|\mathbf{x}_i - \mathbf{x}_j|$ by $\log(|\mathbf{x}_i - \mathbf{x}_j|/\lambda)$, where λ is some minimum separation between vortices; the effect of such a change is to add a constant to H. μ is the self-energy

---

[16] J. M. Kosterlitz and D. Thouless, 1973; J. M. Kosterlitz, 1974; C. Itzykson and J. M. Drouffe, *Statistical Field Theory*, Cambridge, 1989.

of a vortex; for a sparse system, $\mu \sim \frac{\partial H}{\partial N}$; 2μ can be thought of as the energy needed to create a vortex pair of minimum energy. We only wish to consider a system in which the number of positive Γ's equals the number of negative Γ's and $\Sigma \Gamma_i = 0$. The partition function for this system is

$$Z = \int_{\mathcal{D}^N} e^{-\beta H} d\mathbf{x}_1 \cdots d\mathbf{x}_N \; ;$$

$e^{-\beta H} = e^{-N\beta\mu} e^{-\beta H_1}$, where H_1 is the interaction energy. It is convenient to write $e^{-\beta\mu} = K$ and

$$Z = K^N \int_{\mathcal{D}^N} e^{-\beta H_1} d\mathbf{x}_1 \cdots d\mathbf{x}_N \; .$$

Suppose that all pairs of vortices within a distance λ of each other make only a negligible contribution to the Hamiltonian and therefore to Z. We wish to delete all pairs of vortices with separation between λ and $\lambda + \Delta\lambda$ without changing the statistics of the problem, i.e., we wish to renormalize the problem so that it has fewer small-scale structures but its statistics are unchanged. We limit ourselves to the situation where there are few vortices permitted in the given area; μ has to be large. The charge from x to $\lambda + \Delta\lambda$ removes small scales of motion from the system. The difference between the previous spin renormalization and the present renormalization is mostly one of method rather than of principle; the renormalization will change μ, and μ controls N.

If there are no vortex pairs within λ of each other,

$$\int_{\mathcal{D}^N} d\mathbf{x}_1 \cdots d\mathbf{x}_N \equiv \int_{D_N} d\mathbf{x}_N \int_{D_{N-1}} d\mathbf{x}_{N-1} \cdots \int_{D_1} d\mathbf{x}_1 \; ,$$

where D_k is $\mathcal{D}$, the domain to which the vortices have been confined, from which the circles $|\mathbf{x} - \mathbf{x}_j| \leq \lambda$, $j = k+1, k+2, \ldots, N$, have been excised. We wish now to replace the integrand in Z by a new integrand which, when integrated over a domain with larger holes, will produce the same Z. We first have to display in the integration the new annuli of radii $\lambda + \Delta\lambda$ from which vortices will soon be banned, namely,

$$\int_{D^N} d\mathbf{x}_1 \cdots \int_{D_1} d\mathbf{x}_1 \quad = \quad \int_{D'_N} d\mathbf{x}_N \cdots \int_{D'_1} d\mathbf{x}_1$$

$$+ \frac{1}{2} \sum_j \left(\int_{D'_N} d\mathbf{x}_N \cdots \int_{D'_{j-1}} d\mathbf{x}_{j-1} \int_{D'_{j+1}} d\mathbf{x}_{j+1} \right.$$

$$\cdots \int_{D'_{i-1}} d\mathbf{x}_{i-1} \int_{D'_{i+1}} d\mathbf{x}_{i+1} \Big)$$

(6.9)
$$\times \left(\int_{\overline{D}_j} d\mathbf{x}_j \int_{\delta_i(j)} d\mathbf{x}_i \right) ,$$

where D'_k is like D_k but with λ replaced by $\lambda + \Delta\lambda$, $\delta_i(j)$ is the annulus $\lambda < |\mathbf{x}_i - \mathbf{x}_j| < \lambda + \Delta\lambda$, and $\overline{D}_j$ is the plane minus $\delta_i(j)$ and minus circles of radius λ around points other than $\mathbf{x}_i$ and $\mathbf{x}_j$. Areas $O(\Delta\lambda^2)$ have been neglected.

Because $\beta > 0$, we can assume that nearby pairs of vortices of opposing sign contribute much more to Z than pairs of the same sign; it is reasonable to assume that if vortices at $\mathbf{x}_i, \mathbf{x}_j$ are in $\delta_i(j)$, i.e., $|\mathbf{x}_i - \mathbf{x}_j| = \lambda$, then their signs are opposite: $\Gamma_i = -\Gamma_j$. This assumption holds better for $T < T_c$ than for $T > T_c$. Also, $\mathbf{x}_i = \mathbf{x}_j + \boldsymbol{\lambda}$, where $\boldsymbol{\lambda} = (\lambda \cos\theta, \lambda \sin\theta)$ for some angle θ. Carry out the integration of $e^{-\beta H_1}$, H_1 = interaction part of the Hamiltonian, on $\delta_i(j)$, and consider only the interaction of terms that involve either $\mathbf{x}_i$ or $\mathbf{x}_j$:

$$J = \int_{\delta_i(j)} d\mathbf{x}_i \exp\left(\beta \sum_k \Gamma_i\Gamma_k \log|\mathbf{x}_i - \mathbf{x}_k| + \beta \sum_k \Gamma_j\Gamma_k \log|\mathbf{x}_j - \mathbf{x}_k| \right)$$

$$= \int d\mathbf{x}_i \exp\left(\beta\Gamma_i\Gamma_k \left(\log|\mathbf{x}_j + \boldsymbol{\lambda} - \mathbf{x}_k| - \log|\mathbf{x}_i - \mathbf{x}_k|\right) \right)$$

$$= \lambda\Delta\lambda \int_0^{2\pi} d\theta \prod_{\substack{i=1 \\ i \neq j \\ i \neq k}}^N \left(1 + \frac{2\boldsymbol{\lambda} \cdot (\mathbf{x}_j - \mathbf{x}_k)}{|\mathbf{x}_j - \mathbf{x}_k|^2} + \frac{\lambda^2}{|\mathbf{x}_j - \mathbf{x}_k|^2} \right)^{\beta\Gamma_i\Gamma_k}$$

If the vortex "gas" is sparse, then the probability that there is a second vortex in the annulus of radius $\lambda + \Delta\lambda$ around $\mathbf{x}_i$ is small, and $\lambda^2/|\mathbf{x}_j - \mathbf{x}_k|^2$ is small. Expand the integrand of the last integral in powers of $\lambda^2/|\mathbf{x}_j - \mathbf{x}_k|^2$, omit terms $O(\lambda^2)$, integrate over θ, and integrate over $\overline{D}_j$; do the same for all the terms in H_1, and sum. The result is

$$\Sigma J = 2\pi\lambda\Delta\lambda \left(|\mathcal{D}| - 2\pi\lambda^2\beta^2\Gamma^2 \sum_k \sum_{\ell \neq k} \Gamma_k\Gamma_\ell \log|\mathbf{x}_k - \mathbf{x}_\ell| \right)$$

(Note that all dependence on $\mathbf{x}_i, \mathbf{x}_j$ has naturally dropped out.)

Substitution of this last expression into the integration domain formula (6.9), followed by an approximation of the form $1 + B\Delta\lambda \cong \exp(B\Delta\lambda) + O(\Delta\lambda^2)$, yields a new expression for Z:

$$Z = \exp(2\pi K^2 \lambda \Delta\lambda |\mathcal{D}|) K^N \int_{D'_N} d\mathbf{x}_N \cdots \int_{D'_1} d\mathbf{x}_1$$

$$\cdot \exp\left\{ \beta \left(1 - (2\pi)^2 \beta \Gamma^2 (K\lambda^2) \frac{\Delta\lambda}{\lambda} \right) \sum_i \sum_{j \neq i} \Gamma_i \Gamma_j \log |\mathbf{x}_i - \mathbf{x}_j| \right\}.$$

The integration is over annuli of radii $\lambda + \Delta\lambda$. Reintroducing the λ into the log so as to make all the logs non-negative, and changing λ into $\lambda + \Delta\lambda$, we find

$$Z(\beta\Gamma^2, K\lambda^2) = Z_0 Z(\widetilde{\beta\Gamma}^2, \widetilde{K\lambda}^2) \,,$$

where $\widetilde{\beta\Gamma}^2, \widetilde{K\lambda}^2$ are the new, "renormalized" values of these quantities and $Z_0 = \exp(2\pi K^2 \lambda \Delta\lambda |\mathcal{D}|)$. $\widetilde{\beta\Gamma}^2, \widetilde{K\lambda}^2$ are given by

$$\widetilde{\beta\Gamma}^2 = \beta\Gamma^2 \left(1 - (2\pi)^2 (\beta\Gamma^2)(K\lambda^2)^2 \frac{\Delta\lambda}{\lambda} \right)$$

$$\widetilde{K\lambda}^2 = K\lambda^2 \left(1 - (\beta\Gamma^2 - 2) \frac{\Delta\lambda}{\lambda} \right).$$

(The last term, $K\lambda^2 \cdot 2\frac{\Delta\lambda}{\lambda}$ comes from the variation of the factor λ^2 in $\widetilde{K\lambda}^2$ that was introduced for convenience.) If we write $Y = \beta\Gamma^2$, $X = K\lambda^2 = e^{-\beta\mu}\lambda^2$, these equations reduce to

$$(6.10) \qquad \qquad \lambda \frac{dY}{d\lambda} = -(2\pi)^2 X^2 Y^2$$

$$(6.11) \qquad \qquad \lambda \frac{dX}{d\lambda} = -X(Y - 2)$$

Consider the neighborhood of the trivial fixed point $X = 0$, $Y = 2$ (which suggests as a first approximation to T_c the value $T_c = \Gamma^2/2$). Write $\frac{d\lambda}{\lambda} = dq$, $x = X^2$ (thus $x \geq 0$ always), $y = Y - 2$; these equations reduce to

$$\frac{dy}{dq} = -4\pi^2 x (y + 2)^2 \,, \qquad \frac{dx}{dq} = -2xy \;;$$

expand around $x = 0, y = 0$, obtaining

$$\frac{dy}{dq} = -16\pi^2 x \,, \qquad \frac{dx}{dq} = -2xy \,.$$

One can readily see that these equations leave invariant the half-line $y = 8\pi^2 X$, $y \geq 0$, and that this half-line is unstable, in the sense that if X, y start near it, they will eventually get away. This degeneracy is connected with the indefinite starting value λ in the renormalization. Note however the important fact that as λ increases, the "effective" value of $\beta\Gamma^2$ decreases. If β remains fixed, Γ^2 decreases. Since $\beta > 0$, low-energy configurations are favored; given a vortex pair with large separation, other, smaller pairs will arrange themselves so as to reduce the energy, thus decreasing the "effective" interaction of the vortices in the large pair. If small pairs are removed, one has to make up for the removal by reducing Γ. If one compares this conclusion with the discussion in Section 3.1, one sees that for $T > 0$ the "large scales" (pairs with large separation) and "small scales" (pairs with small separations) are not statistically independent: the former rearrange the latter, the latter attenuate the former.

6.8. Vortex Percolation/λ Transition in Three Space Dimensions

We now generalize the construction of the preceding section to three space dimensions. Consider a collection of small vortex loops in three dimensions (i.e., Buttke loops, described by their magnetization vector $\mathbf{M}$). The Hamiltonian is given by equation (1.28). At low temperatures there will be a few such loops in a finite volume. As T increases, there will be more and more. New loops can arrange themselves so as to reduce the energy associated with the loops already there. The total vorticity in each loop adds up to zero; the system is analogous to a system of plane vortices with $\Sigma\Gamma_i = 0$.

Suppose there is a minimum vortex circulation and a minimum radius associated with each loop. Their values can be explained by quantum mechanics, and/or by placing them on a lattice. The Hamiltonian acquires then a term of the form $N\mu$, where N is the number of loops and μ is the cost in energy of creating a loop. A renormalization procedure, exactly analogous to the procedure of the preceding section, produces a "flow" in parameter space that renormalizes the system.[17] The fixed points of this flow has properties quite different from those of the two-dimensional system of the preceding section. As λ, the minimum loop site, increases, the circulation in the loops must decrease if $\beta > 0$, for the same reason as in two dimensions: loops arrange each other so as to decrease the energy, because the Gibbs factor favors low energies.

It has been suggested that the critical point of this system can be viewed as a percolation threshold in a correlated percolation problem. "Correlated" percolation is percolation in a system in which the probabilities of

---

[17]See, e.g., G. Williams, 1987.

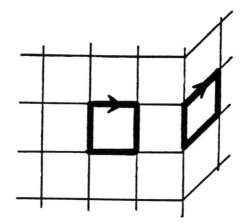

FIGURE 6.7. Elementary vortex loops on a lattice.

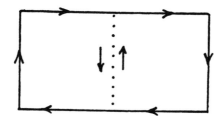

FIGURE 6.8. Two adjoining loops.

events in different parts of a lattice are not independent. Assume the loops live on a three-dimensional cubic lattice.

Place, or fail to place, on each square facet (= "plaquette") of a unit cube an elementary vortex loop (Figure 6.7). There are six types of elementary loops: in a plane parallel to the (x_1, x_2) plane, in a plane parallel to the (x_2, x_3) plane, in a plane parallel to the (x_1, x_3) plane, and in each case, oriented in one of two ways. Two adjoining elementary loops with the same orientation have a bond that cancels, and the result is a larger, "macroscopic" vortex loop that traces out the boundary of the union of the two elementary loops (Figure 6.8).

As T and the number of elementary loops increases, there appear longer macroscopic vortex filaments[18] that trace out the boundaries of percolation clusters of vortex loops that live on the faces of the cubic lattice.[19] It is plausible (but not well verified) that at the critical temperature of the lattice loop problem there appear vortex loops of the type just described

---

[18] J. Epiney, 1990.
[19] A. Chorin, 1992.

that have arbitrary length, creating a percolation threshold for a "vortex" percolation problem.

It has been suggested[20] that the unstable fixed point of this loop problem corresponds to the "λ-transition" (so-called because of the shape of its phase diagram) between superfluid and ordinary fluid ^{4}He. The analogy further suggests[21] that the fractal dimension of vortex filaments at T_c is related to that of polymers. This suggestion is very attractive, both because of the analogy to the two-dimensional case and for its intrinsic plausibility. In particular, since viscosity at a boundary slows down fluid by creating vortices in it,[22] superfluid should cease to be a superfluid when it can support macroscopic vortices. This identification of a percolation threshold for vortex filaments and the λ-transition is, however, controversial.

---

[20] G. Williams, 1987; S. Shenoy, 1989.

[21] S. Shenoy, 1989.

[22] A. Chorin, 1973.

7
Vortex Equilibria in Three-Dimensional Space

The tools of the preceding section are used to formulate and solve a simplified model of the equilibrium states of three-dimensional vortex filaments. The connection with the Kolmogorov theory is made, and the implications of the theory for superfluidity and for the numerical modelling of turbulence are discussed.

7.1. A Vortex Filament Model

We formulate a simplified vortex model[1] that can be used to analyze further the phenomena described in Chapter 5, taking advantage of the tools in Chapter 6. The model consists of a sparse suspension of self-avoiding vortex filaments, in a Gibbs ensemble based on the three-dimensional vortex energy function. In the early stages of the analysis the vortex will be confined to a lattice and have a fixed finite length; these constraints will be removed in due time.

Consider an ensemble of vortices supported by N-step oriented SAW's, with each configuration C having a probability $P(C) = Z^{-1}\exp(-\beta E)$, Z = partition function and $E = E(C) = (8\pi)^{-1}\sum_I \sum_{J\neq I} \mathbf{\Gamma}_I \cdot \mathbf{\Gamma}_J/|I-J|$, with $|\mathbf{\Gamma}_I|$ the circulation of the vortex, I a multi-index specifying a location on the lattice, and $|I-J|$ the distance between $\mathbf{\Gamma}_I$ and $\mathbf{\Gamma}_J$, as in Section 5.5. Self-avoidance guarantees $I \neq J$, and thus E is finite. Guided by

---

[1] A. Chorin, 1991; A. Chorin and J. Akao, 1991.

the two-dimensional analysis, we consider both $T < 0$ and $T > 0$. As long as N is fixed, one can neglect the nearly constant term $\mu'N$ in the energy; a factor common to all the configurations cancels out after division by Z. The filament is part of a sparse homogeneous suspension, with the other filaments far enough so that they do not reduce the number of configurations available to the filament at hand. The interaction between different filaments is accounted for in the use of the canonical ensemble. The limit $N \to \infty$ will be taken after the finite N case is discussed.

The vortex filaments can be either open or closed. The conclusions are independent of this choice. Closed filaments may be more attractive since div $\boldsymbol{\xi} = 0$; on the other hand, as a result of intermittency, the more concentrated portions of a vortex filament may end in less concentrated filaments and the large k range we shall be investigating may be better represented by a union of open filaments.

The model has two major flaws:

I. The assumption that the vortex suspension is sparse fails to describe the full range of interactions in a vortical flow. As was shown in the two-dimensional case, all the elements of the vorticity field interact strongly. Here, only the pieces of vorticity that belong to a single filament interact strongly while different filaments interact weakly, acting as a heat bath for each other. The features of the problem that mitigate this flaw are: the interaction in three space dimensions has a shorter range than in two dimensions; experience in two dimensions (in particular in connection with the number of imposed invariants) shows that imperfect models may give adequate solutions, and, most importantly, a numerical method that allows one to gauge the validity of the conclusions in the non-sparse case will be described; it affirms the conclusions.

II. The model fails to represent the extraordinary complexity of vortex cross-sections. Some of that complexity will have to be reintroduced at a later stage, in the calculation of inertial exponents. This initial neglect of the statistics of the cross-sections is in fact the model's worst flaw.

On the plus side, the model will afford a relatively simple analysis and lead to useful conclusions. Some of the salient qualitative features of vortex dynamics, in particular the close association of stretching and folding, are preserved in a natural way.

Note that if $|T| = \infty$, $P(C)$ is a constant independent of C, and the filament is a polymer (equal-probability SAW). When $|T| = \infty$, $\frac{dS}{dE} = T^{-1} = 0$ (N fixed); one can readily see that S is decreasing on both sides of the $|T| = \infty$ point, and therefore S has a maximum at the $|T| = \infty$ "polymeric" point. The vortex equilibrium of maximum entropy is a polymer.

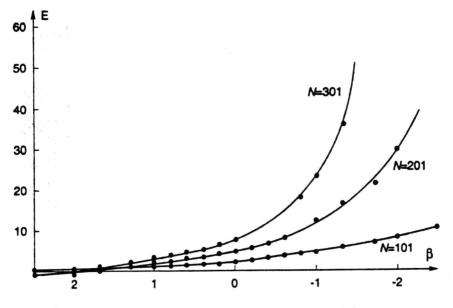

FIGURE 7.1. Mean energy as a function of β.

7.2. Self-Avoiding Filaments of Finite Length

The properties of an N-step filament can be calculated numerically, by constructing an appropriate Markov chain of filament configurations. The calculations verify that, as always, the mean energy $\langle E \rangle$ ($= E$ for short when no ambiguity arises) is an increasing function of the temperature. The variation of E with T is plotted in Figure 7.1 (for convenience, $\Gamma = |\Gamma| = 1$).[2] The values of $\beta = T^{-1}$ are arranged so that the temperature increases as one moves to the right (see Chapter 4).

The mean energy of the filament increases with N (Figure 7.2).

The entropy per vortex leg is $S/N = -\sum_C P(C) \log P(C)/N$; the probability $P(C)$ can be evaluated by estimating the frequency of occurrence of a configuration C in the appropriate Markov chain.[3] The entropy per leg S/N is plotted in Figure 7.3 as a function of β. As advertised, S is maximum when $\beta = 0$ ($|T| = \infty$). The curve is steeper for large T because E is larger for large T and $T^{-1} = \frac{dS}{dE}$.

One can also verify that S increases with N—the longer the vortex, the larger its entropy. This is the promised converse of the analysis in Section 5.1, where it was shown that an increase in entropy leads to an increase in vortex length. When $|T| = \infty$, one can conclude that the entropy of a

---

[2]A. Chorin, 1991.
[3]H. Meirovitch, 1984; A. Chorin and J. Akao, 1991.

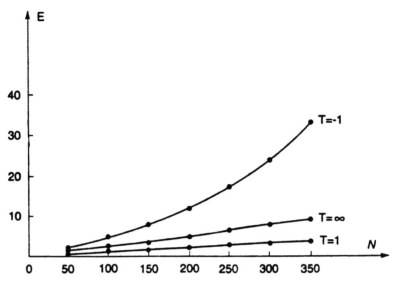

FIGURE 7.2. Mean energy of a filament as a function of N.

polymer increases with its length—a well-known fact.[4]

Given a configuration of a vortex filament with N legs, one can calculate the straight-line distance r_N between the first and last leg, assuming the filament is open, and the function $\mathcal{S}_r = \langle \sum_{|I-J|\leq r} \boldsymbol{\Gamma}_I \cdot \boldsymbol{\Gamma}_J \rangle$ that enters the definition of the vector-vector correlation exponent. One can then calculate

$$\mu_1 = \mu_{1,N} = \mu_1(N,T) \quad = \quad \frac{\log \langle r_N^2 \rangle^{1/2}}{\log N} \, ,$$

$$\mu_2(N,T,r) \quad = \quad \frac{\log r}{\log \mathcal{S}_r} \, .$$

$\mu_2(N,T,r)$ is well defined for r small enough so that the vortex has a small probability of ending within the sphere $|I - J| \leq r$. A good estimate of the acceptable r is given by $\langle r_N^2 \rangle^{1/2} = \bar{r}_N$; we therefore define

$$\mu_{2,N} = \mu_2(N,T) = \frac{\log \bar{r}_N}{\log \mathcal{S}_N} \, ,$$

where $\bar{r}_N$ has just been defined and $\mathcal{S}_N = \mathcal{S}_{\bar{r}_N}$. If $\lim_{N \to \infty} \mu_{1,N}$ exists and equal μ_1, then μ_1 is an analogue of the Flory exponent and $D_1 = 1/\mu_1$ is the fractal dimension of the filament. Similarly, if $\mu_{2,N} \to \tilde{\mu}$, the energy spectrum is $E(k) \sim k^{-\tilde{D}}$, $\tilde{D} = 1/\tilde{\mu}$. $\mu_{1,N}$, $\mu_{2,N}$ as functions of $\beta = 1/T$ and of N are plotted in Figures 7.4 and 7.5.

---

[4]P.G. de Gennes, *loc. cit.*

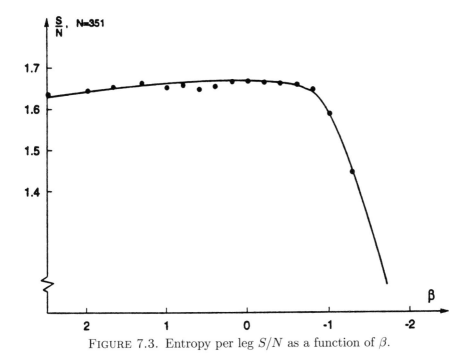

FIGURE 7.3. Entropy per leg S/N as a function of β.

The graph of $\mu_{1,N}$ as a function of N and β shows that for fixed N, $D_N = 1/\mu_{1,N}$ decreases as T increases (remembering that $T < 0$ is "hotter" than $T > 0$). This can be understood in the framework of the two-dimensional theory of Chapter 4 and of the Kosterlitz-Thouless theory of Chapter 6: Cut the filament by a plane; at the points of intersection, the direction of the vortex alternates; the set of intersection points looks like a collection of point vortices of alternating sign. The distance between vortices of opposing sign cannot increase and lead to segregation into vortex patches of differing signs, as in Chapter 4, because the intersections lie on a three-dimensional object. An increase in T can only express itself as an increase in the separation r between point vortices of opposing sign, and therefore is an unfolding of the vortex and a decrease in D_N. Furthermore, if the energy of the vortex is fixed, an increase in N must bring a decrease in T and then an increase in D; in other words, as a vortex filament is stretched, it must fold.

In all these graphs, the vortex filament has a constant circulation (for the sake of definiteness, $\Gamma = 1$). Contrast two vortex filaments with the same finite N but different circulations Γ_1, Γ_2, say $\Gamma_1 > \Gamma_2$. The energy integral being proportional to Γ^2, the Gibbs weights attached to the two filaments are $Z^{-1} \exp(-\beta \Gamma_1^2 E)$, $Z^{-1} \exp(-\beta \Gamma_2^2 E)$, where E is the energy that results from $\Gamma = 1$. These weights are the same as those one would obtain with

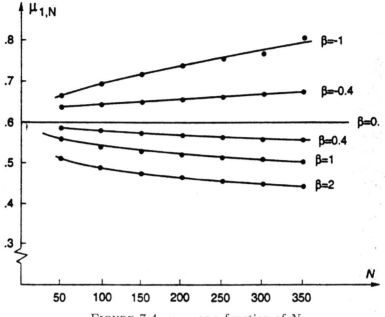

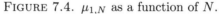

FIGURE 7.4. $\mu_{1,N}$ as a function of N.

$\Gamma = 1$ and $T_1 = T/\Gamma_1^2$ in the first case, $T_2 = T/\Gamma_2^2$ in the second. If one thinks of $D_{1,N} = 1/\mu_{1,N}$ as an approximate fractal dimension, the vortex with larger Γ has a smaller $|T|$, and if $\beta < 0$ (which we shall see is the only physically relevant case), then the vortex with larger Γ has a smaller $D_{1,N}$ and appears smoother. Strong vortices are less folded; this has been observed numerically.[5] Furthermore, a collection of vortex filaments with finite N looks multifractal, in the sense of Chapter 3.

7.3. The Limit $N \to \infty$ and the Kolmogorov Exponent

Assume the trends in Figures 7.4 and 7.5 continue as $N \to \infty$. The conclusion one reaches is:

$$\lim_{N \to \infty} \mu_{1,N} = \begin{cases} 1 & T < 0 \\ \mu & |T| = \infty \\ 1/3 & T > 0 \end{cases}$$

where $\mu \cong 0.588$ is the Flory exponent. 1 is the upper bound on the possible values of μ and $1/3$ is the lower bound (no object in the three-dimensional space can have dimension $D = 1/\mu$ larger than 3). When the limit is 1 the

---

[5]Z. S. She, personal communication.

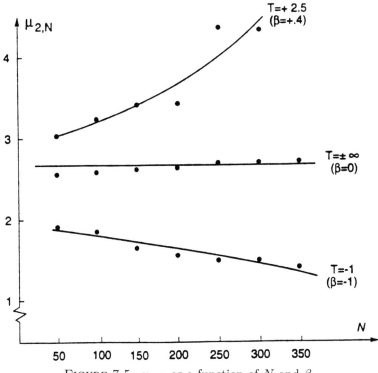

FIGURE 7.5. $\mu_{2,N}$ as a function of N and β.

filament is smooth, and when the limit is $1/3$ the filament is as balled-up as the space available would allow.

Similarly,

$$
\lim_{N\to\infty} \mu_{2,N} = \begin{cases} 1 & T < 0 \\ \tilde{\mu} & |T| = \infty \\ \infty & T > 0, \end{cases}
$$

where the limits 1 and ∞ correspond to the upper and lower bounds on $\tilde{D} = 1/\tilde{\mu}$: $0 \le \tilde{D} \le 1$ (see Section 6.3). In the same section, we gave the estimates $\tilde{D} \cong 0.37$ (numerical), $\tilde{D} = 0.25 + O(1)$ (analytical). If we knew nothing further about vortex filaments, we would conclude that for $T < 0$, $D = \dim \mathrm{supp}\, \boldsymbol{\xi} = 1$, and $\gamma = 1$ [$E(k) \sim k^{-\gamma}$ for large k]; for $T > 0$, we would obtain $D = \dim \mathrm{supp}\, \boldsymbol{\xi} = 3$, $\gamma = 0$, and for $|T| = \infty$, $D \cong 1.70$, $\gamma \cong 0.37$. However, we know that the filament model is an idealization, the vortex has a non-trivial cross-section, and $D > D_c$, where D_c is the fractal dimension of the centerline of the vortex. We now present a recipe for taking the cross-section into account. A discussion of the shortcomings and the relevance of the recipe will follow.

Consider a vortex tube with a non-trivial cross-section, forming part of a homogeneous sparse suspension of such tubes. Take a point x in the tube, and evaluate the integral

$$S_r = \left\langle \int_{|\mathbf{r}| \leq r} \boldsymbol{\xi}(\mathbf{x}) \cdot \boldsymbol{\xi}(\mathbf{x} + \mathbf{r}) d\mathbf{r} \right\rangle;$$

this integral is the continuum version of the sum used to define the vector-vector exponent for a polymer. Suppose the tube is thin, has centerline C and cross-section Σ, and suppose r is small enough so that there are no other tubes within r of $\mathbf{x}$. Then

$$S_r = \left\langle \boldsymbol{\xi}(\mathbf{x}) \cdot \left[\int_{C_s} ds \int_{\Sigma(s)} \boldsymbol{\xi}(\mathbf{x} + \mathbf{r}) d\Sigma(s) \right] \right\rangle,$$

where $\Sigma(s)$ is a cross-section of the tube, at a distance s from $\mathbf{x}$ measured along the centerline, $\mathbf{r} = \mathbf{x}(s) - \mathbf{x}$ is the vector from $\Sigma(s)$ to $\mathbf{x}$, and C_s is the portion of the centerline within r of $\mathbf{x}$. C_s may consist of several unconnected portions. If $\boldsymbol{\xi}$ is distributed uniformly on $\Sigma(s)$ and $\Gamma = 1$, then

$$S_r = \left\langle \boldsymbol{\xi}(\mathbf{x}) \cdot \int_{C_s} |\Sigma| \boldsymbol{\xi}(\mathbf{x} + \mathbf{r}) ds \right\rangle,$$

where $|\Sigma|$ is a characterization of the size of Σ. If Σ is fractal, $|\Sigma| \sim r^{D_\Sigma}$, where D_Σ is $\dim \Sigma$. If $\dim \text{supp } \boldsymbol{\xi} = D$, then $\dim \Sigma = D - D_c$, where D_c is the dimension of the centerline of the tube. If $\langle \int_{C_s} \boldsymbol{\xi}(\mathbf{x}) \cdot \boldsymbol{\xi}(\mathbf{x} + \mathbf{r}) ds \rangle \sim r^{\tilde{D}}$, the contribution of the sphere $|\mathbf{r}| \leq r$ to the correlation function is $\sim r^{D - D_c + \tilde{D}}$, and following the usual analysis, $E(k) \sim k^{-\gamma}$, with

(7.1) $\gamma = D - D_c + \tilde{D};$

where $D = \dim \text{supp } \boldsymbol{\xi}$, $D_c = $ dimension of the centerline, $\tilde{D} = 1/\tilde{\mu} = $ vector-vector correlation exponent. In particular, if $|T| = \infty$, $D \sim 3$, then

$$\gamma \cong 3 - 1.70 + 0.37 = 1.67 \cong 5/3.$$

The Kolmogorov exponent has been recovered in the $|T| = \infty$ case.

Here are some of the things that are unsatisfactory about this argument:

(i) The main problem is that the correction for the presence of a non-trivial cross-section is ad hoc. We obtained $D_c \cong 1.70$ and $\tilde{D} \cong 0.37$ for a thin filament with trivial cross-section. The extreme variability of a vortex cross-section makes plausible the idea that cross-sections can be tacked on as an afterthought, but the conclusion is not certain. There has to be an accounting for cross-sections in a complete model of turbulence, but the one just given in incomplete.

(ii) If dim supp $\boldsymbol{\xi} = D > D_c$, the "cross-section" of the vortex must have dimension $D - D_c$, or else the rest of the support contained in a sphere of radius r will not grow like r^D. However, a cross-section that grows as r grows is not localized (i.e., not confined to a tube of bounded radius), and this emphasizes the difficulty in viewing a vortex tube as a thin filament.

(iii) The values $D \sim 3$ and $\tilde{D} \cong 0.37$ are uncertain; further work may change them. The close agreement between $D - D_c + \tilde{D}$ and the Kolmogorov exponent $5/3$ is probably fortuitous. In view of i) and ii), this is only to be expected.

Note that if the Kolmogorov law is to hold, it is necessary that Σ be fractal. Suppose to the contrary that Σ can somehow be identified with a small area $(D = 2)$ of radius σ. For $r < \sigma$ one readily sees that $S_r \sim r^3$, and thus $E(k) = 0$ for $k \ll \sigma^{-1}$; for $r \gg \sigma$, Σ does not affect the exponent. Thus a finite, non-fractal cross-section simply chops off the spectrum and does not produce a self-similar spectrum.

A positive aspect of the calculation is the fact that numerical calculations[6] show that a stretching, folding, fractalizing and sheetifying vortex does produce a Kolmogorov spectrum.

For $T < 0$, formula (7.1) yields $\gamma = 3 - 1 + 1 = 3$, and $E(k) \sim k^{-3}$. If $\mu = 1$, $\tilde{\mu} = 1$, the vortex lines are smooth and presumably so is the flow. As shown in Chapter 3, $\gamma = 3^+$, i.e., $E(k)$ decays faster than k^{-3} for large k. The value $\gamma = 3$ emerges as a consequence of the filament assumption, and is an underestimate of γ. For $T > 0$, (7.1) gives $\gamma = 0$. We shall see below that in this case too the filament assumption fails because reconnections destroy the filament. The filament model holds only for $|T| = \infty$, which is fortunately the only case of real interest.

Relations between exponents are often dimension-independent. In a space of dimension one or two, equation (7.1) is not meaningful because there are no vortex filaments. Consider what happens to (7.1) in a space of dimension $d \geq 4$. Note first that (7.1) cannot be derived directly when $d \geq 4$ because a polymer in $d \geq 4$ is indistinguishable from a Brownian walk, the vector-vector correlation is 0 ($\tilde{D} = 0$), and remains 0 after multiplication by r^{D-D_c}, thus $\gamma = 0$; $\gamma = 0$ is one possible generalization of (7.1) to $d \geq 4$. On the other hand, suppose (7.1) is assumed to hold for $d \geq 4$ by continuation from $d = 3$. We have seen that $d = 4$ is the upper critical dimension for the polymer problem; four-dimensional space is wide enough so that self-avoidance does not constrain a walk, and $\mu = 1/2$, $D_c = 2$, $\tilde{D} = 0$.

---

[6] A. Chorin, 1981; J. Bell and D. Marcus, 1992.

Suppose $d = 4$ is also the upper critical dimension for an energy-conserving vortex filament, i.e., that for $d \geq 4$ energy conservation no longer restricts significantly the possible configurations of a vortex filament. The movement of energy from scale to scale is unhindered, the average time energy spends in a scale ℓ_n is $\tau_n = \ell_n/u_n$, $u_n = \sqrt{E}$, $E =$ energy; this energy can move up or down in scale. The corresponding spectrum is $E(k) \sim k^{-2}$ (Chapter 3), $\gamma = 2$, and (7.1) is verified for $d = 4$: $2 = 4 - 2 + 0$. Thus $\gamma = 2$ is a reasonable candidate for the exponent in a "mean-field" theory, which is what one usually obtains at the upper critical dimension, and $\gamma = 5/3$ is what one gets after an "intermittency correction", which results from the constraints on vortex configurations that are due to energy conservation in the stretching process. These remarks were already made in Chapter 3.

Note that we have obtained $\gamma = 5/3$ with $D = 3$. One can use (7.1) to define a family of models with $\gamma = \gamma(D)$, as described in Chapter 3. The sign of $\frac{d\gamma}{dD}$ is unknown a priori, because $\frac{dD_c}{dD}$, $\frac{d\tilde{D}}{dD}$ are unknown. The argument of the preceding paragraph suggests $\frac{d\gamma}{dD} > 0$, as is the case for exponents that relate to scalar quantities (section 6.2).

7.4. Dynamics of a Vortex Filament: Viscosity and Reconnection

Having examined the equilibria of vortex filaments, what can we conclude about the dynamics of vortex filaments?

Suppose at $t = 0$ one is given a sparse suspension of smooth vortex filaments. Smoothness implies $T < 0$; smooth vortex filaments can be approximated by a finite union of straight vortex tubes of finite length, as in Section 5.2, and they behave as if N were finite. The Euler and the Navier-Stokes equations cause filaments to stretch and fold, and N increases. The energy is an increasing function of both T and N; if energy is conserved and N increases T must decrease. If $T < 0$, T decreases as $|T|$ increases. As long as $T < 0$, an increase in N brings one closer to the $|T| = \infty$ boundary between $T < 0$ and $T > 0$, where the entropy is maximum for a given N and where the Kolmogorov spectrum reigns. This variation in T does not contradict the equilibrium assumption as long as it is gradual enough for the corresponding states to be viewed as a succession of equilibria.

Note that the situation is analogous to what happens to a perfect gas when the number of molecules is increased without an energy increase (Section 4.1). The situation is different from the two-dimensional vortex case, where T depends on N but the physical results are independent of N; the difference lies in the fact that in two dimensions N is fixed as the system evolves .

If N is large enough, the $|T| = \infty$ point is a barrier that cannot be crossed because:

(i) General principles: $|T| = \infty$ is the maximum entropy state (but a note of caution is needed: it is the maximum entropy state for fixed N).

(ii) Energy conservation: the sudden folding, from $D_c = 1$ to $D_c = 3$, that accompanies a change in the sign of T, cannot be accommodated if the system conserves energy; an increase in D_c should sharply decrease the energy.

The limit $N \to \infty$ requires caution: As $N \to \infty$, either with the lattice spacing remaining fixed, or with the lattice spacing $h \to 0$, the energy increases to infinity (this has been shown when it was shown that $\tilde{D} \leq 1$). The limit $N \to \infty$ with $h \to 0$ is the interesting limit, because if h is fixed and $N \to \infty$ the size of the system increases, and the increase in energy is natural. $h \to 0$ when smaller and smaller scales come into play. As $h \to 0$, the lattice cut-off implicit in the energy formula

$$E = \frac{1}{8\pi} \sum_I \sum_{J \neq I} \mathbf{\Gamma}_I \cdot \mathbf{\Gamma}_J / |I - J|$$

becomes ineffective.

The easiest way to deal with this limit is to reintroduce a chemical potential $\mu' N$ (we shall from now on drop the prime); the term μN was dropped when the filament with finite fixed N was discussed. One can cover the filament by a union of finite, overlapping balls, and relegate the interactions of the bonds within a ball to a chemical potential term. The interaction energy is now finite, and remains finite as $N \to \infty$ while the radii of the balls remain bounded from below. This is a simplified way to dealing with the non-trivial cross-sections of the vortex filament. The picture we are presenting is self-consistent if for every finite ball radius, the limit $N \to \infty$ brings one to the neighborhood of the $|T| = \infty$ point. The use of finite ball radii was implicit in the renormalization calculation of Section 6.7, where the renormalization was started with λ finite. Indeed, the limiting process $N \to \infty$ is the inverse of the renormalizations of Sections 6.7 and 6.8; the line $|T| = \infty$, in a (T, λ) parameter space, should be attracting as $N \to \infty$ for a finite λ; the point $|T| = \infty$, $\lambda = 0$ should be a critical point. Thus the attracting (stable) point for physical flow is the unstable point of the renormalization flow. The analysis of this situation is at present incomplete.

This analysis suggests that the equilibrium at $|T| = \infty$ is attracting, at least for Euler flow and an initially smooth vortex filament ($T < 0$). The attracting equilibrium has a Kolmogorov spectrum, according to the calculations of the preceding section. The statistics of the velocity field that result from this vortex equilibrium are not Gaussian; indeed the Biot-Savart formula gives $\mathbf{u}$ as a limit of expressions of the form

$$\mathbf{u}(\mathbf{y}) = \Gamma \int_{\text{filament}} K\left(\mathbf{y} - \mathbf{x}(s,\omega)\right) ds,$$

where Γ is a circulation, K is the Biot-Savart kernel, and $\mathbf{x} = \mathbf{x}(s,\omega)$ is a random vortex filament. Such integrals are known not to be Gaussian.[7] Some numerical calculations[8] have yielded values of the flatness of $\frac{\partial u}{\partial x}$ that are greater than 3. The skewness is of course zero as long as we have not introduced a viscosity because we are dealing with a thermal equilibrium, in which each orientation of the vortex filament is equally likely.

Indeed, if the viscosity ν is not zero, the increase in N is halted, because fractalization cannot proceed beyond the Kolmogorov scale η and therefore there can be no convergence to a state with a self-similar spectrum that extends all the way to infinity. With a finite N, the $|T| = \infty$ threshold can be crossed. In Figure 7.6 we exhibit an example of the evolution of two filaments with different energies E on a lattice with a fixed lattice spacing. The larger E corresponds to a larger initial N. In the case of a larger N the asymptote $|T| = \infty$ is respected, in the case of a smaller N it is not. This can happen because for finite N the maximum entropy principle is less compelling and the drop in E at $|T| = \infty$ less steep; the fluctuation in E allowed by the canonical ensemble permits the jump. These observations on the effect of a finite viscosity are consistent with the observation that in two dimensions a finite viscosity lowers a negative temperature towards the positive temperature domain.

Consider the effect of the crossing of the $|T| = \infty$ barrier at finite ν. As T decreases, D_N, the approximate fractal dimension of the filament, increases, and the number of close encounters between counterrotating vortex legs increases. Such encounters lead to cancellations, reconnections, and a subsequent decrease in D_N. It is a plausible conjecture that the exponents $\mu, \tilde{\mu}$ are kept at their equilibrium value by reconnection. Thus the effect of viscosity and the consequent energy loss manifest themselves in the shedding of vortex loops by vortex filaments[9], maintaining the collection of vortex filaments in equilibrium. An accessible illustration of this mechanism is present in aircraft trailing vortices which bend and stretch, thus decreasing T towards infinity. When the two trailing vortices touch they always reconnect, since, as we have just shown, the reconnected state has higher entropy.

---

[7] A. Chorin, 1990.

[8] A. Chorin and J. Akao, 1991.

[9] A. Chorin, 1991; A. Bershadski and A. Tsinober, 1991.

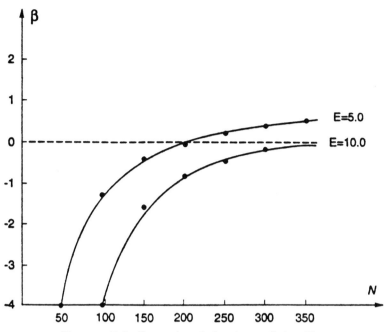

FIGURE 7.6. Long time behavior at finite N.

The shedding hypothesis resembles the "self-organized criticality" model of critical phenomena.[10] A sand-pile is a paradigm of self-organized criticality: As one pours sand on top of it, its slope will stabilize near a critical value where any perturbation can cause a catastrophic avalanche. The shedding of loops is an avalanche-like event that maintains the slope of the energy spectrum $E(k) \sim k^{-5/3}$. For $T > 0$ the slope would be flatter than that of $k^{-5/3}$, for $T < 0$ it is steeper.

Consider for a moment closed rather than open vortex loops. If one replaces a vortex filament by the equivalent collection of elementary (Buttke) loops that it spans (see Section 6.8), then the physical vortex is the boundary of clusters of elementary loops. The shedding of a closed vortex loop by a given loop is a decrease in the size of the enclosed cluster. This picture is consistent with the idea that at $|T| = \infty$ the size of the cluster is maximum, and thus it is plausible that the macroscopic vortices at $|T| = \infty$ coincide with the hulls of clusters of elementary Buttke vortex loops at a percolation threshold, just as was suggested for the λ-transition in Section 6.8; the $|T| = \infty$ state is the intermediate state between the smooth vortices at $T < 0$ and the very folded vortices at $T > 0$. An analogous identity between an intermediate state for folding polymers and percolation cluster

---

[10]P. Bak et al., 1988.

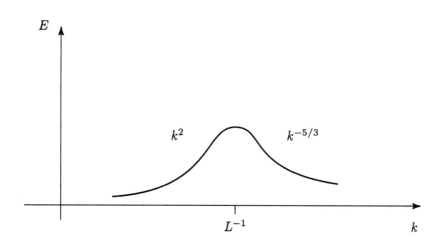

FIGURE 7.7. A spectrum with two kinds of equilibrium.

hulls was presented in Section 6.5.

If one succeeds in maintaining $T > 0$ with a small but finite viscosity, reconnections should destroy all the large loops; the constraints imposed by the connectivity of vortex lines and the conservation of circulation become inoperative. In the absence of constraints the equilibrium should coincide with the "absolute" equilibrium of Section 4.2, with $E(k) \sim k^2$. The calculation of γ based on a filament model fails when the filaments cease to be large. More generally, if one has a flow where vortex loops of diameter L are produced, L smaller than the dimensions of the fluid container, one expects that for $k \ll L^{-1}$ the connectivity and circulation constraints are unimportant and $E(k) \sim k^2$, while for $k > L^{-1}$ these constraints are important and $E(k) \sim k^{-5/3}$, resulting in spectra of the form in Figure 7.7. Such spectra are produced in certain geophysical flows.[11]

According to the theory presented in this section, the universal equilibrium of Chapter 3 is a statistical equilibrium in the usual sense. The small and the large scales are strongly coupled; without the small scales the convergence to the $|T| = \infty$ state does not happen. When $|T| = \infty$, a large vortex loop does not polarize smaller ones (see Sections 6.7 and 6.8), nor does it "antipolarize" them by weighting more heavily large energy states, as happens when $T < 0$. Thus the small and the large scales, though coupled, are statistically independent. In the case of Euler flow, the $|T| = \infty$ state is the only statistical equilibrium that allows an infinite enstrophy, thus answering the question at the end of Chapter 5.

---

[11]See, e.g., U. Frisch and G. Parisi, 1985; note that these authors label the Kolmogorov range as not being an equilibrium.

Viscosity does cause the equilibrium to deviate slightly from equilibrium, probably in spurts localized in time and space. One can think of the scales at equilibrium as constituting a large lake in equilibrium (a lake now replacing a bathtub in the imagery) with a source at one end and a sink at the other, adding and subtracting energy without significantly affecting the equilibrium, except in occasional bursts.

7.5. Relation to the λ Transition in Superfluids: Denser Suspensions of Vortices

The description[12] of the $|T| = \infty$ state as a percolation threshold resembles the description of the λ transition in Section 6.8; the temperature in the λ transition is positive and small, and one may well wonder whether there is a connection between these two thresholds. We shall now present a simplified model that exhibits analogues of both transitions and a clear connection between them. This model also suggests ways of extending the analysis of a vortex filament system to denser suspensions of filaments. Some of the relations between turbulence and superfluid vortex dynamics will also become clearer.

Consider a two-dimensional square lattice; an elementary vortex coincides with the sides of a lattice plaquette, as in Section 6.8. The elementary vortex is always oriented clockwise. As in Section 6.8, if two adjoining plaquettes support elementary vortices, the vortex lines on their common bond cancel and one obtains a larger, "macroscopic" vortex that coincides with the boundary of their union. It is important to note that even though the vorticity field thus constructed lies in a plane, the resulting velocity field is three-dimensional. Assume first that one goes from plaquette to plaquette on the lattice and with probability p one places an elementary vortex on a plaquette and with probability $1 - p$ one leaves the plaquette empty, each decision being independent of the previous ones. Trace out the resulting macroscopic vortex loops, which are the boundaries of clusters of occupied plaquettes. Note that if an empty plaquette is surrounded by occupied plaquettes the resulting macroscopic vortex is oriented anticlockwise; even though the elementary vortices are all oriented one way, the macroscopic ones can be oriented either way.

If (i, j) are the coordinates of the center of a plaquette, and if there are vortex loops at $(i, j), (i + 1, j + 1)$ but not at $(i, j + 1), (i + 1, j)$, it is ambiguous whether the elementary vortex loops at $(i, j), (i + 1, j + 1)$ would be viewed as connected or not. We adopt the following convention: if $(i + j)$ is even, elementary vortex loops at (i, j) and $(i + 1, j + 1)$, or at (i, j) and $(i - 1, j - 1)$ are viewed as connected (i.e., a loop is connected

[12] A. Chorin, 1992.

to the northeast and southwest); if $(i + j)$ is odd, loops are connected to the southeast and northwest. This convention leads to the site percolation problem of Figure 6.5, where the percolation threshold is $p = p_c = 1/2$.

An arbitrarily long macroscopic vortex can exist only if there can be clusters of elementary vortex loops of arbitrary size, i.e., if $p \leq p_c$; the complement of the set of occupied plaquettes must also contain a connected set of arbitrary size, i.e., $p \geq p_c$. An arbitrarily long macroscopic vortex can thus exist only if $p = p_c = 1/2$. (At $p = p_c$, there does not necessarily exist a vortex of infinite length[13]). We have chosen the rules of connection so as to create a single value of $p = p_c$ at which an arbitrarily large macroscopic vortex can exist. We shall call p_c the "vortex percolation threshold". The fractal dimension of that long macroscopic vortex is 7/4, as discussed in Chapter 6.

The critical probability p_c can be expressed in terms of a temperature. Let $\mu > 0$ be the chemical potential of a vortex loop, i.e., the cost in energy of creating an elementary loop. Each plaquette has two states, one with a vortex loop and one without; the probability of a plaquette being occupied is $p = e^{-\beta\mu}/(1 + e^{-\beta\mu})$, $\beta = 1/T$; $p = 1/2$ corresponds to $|T| = \infty$, $\beta = 0$. Thus percolation occurs at $|T| = \infty$.

Consider now a simple model in which the percolation is correlated, i.e., the probability of a plaquette being occupied is not independent of what is happening around it. Consider a 3×3 block of plaquettes, with energy

$$E = \sum_I \sum_{J \neq I} \frac{\mathbf{\Gamma}_I \cdot \mathbf{\Gamma}_J}{|I - J|} + \ell\mu = E_1 + \ell\mu,$$

where $\mathbf{\Gamma}_I, \mathbf{\Gamma}_J$ are vectors placed on the vortex legs and pointing in a direction consistent with their orientation, ℓ is the number of legs in the macroscopic vortex (i.e., not taking into account legs that have cancelled when neighboring plaquettes contain elementary vortex loops); μ is now the chemical potential of a vortex leg rather than that of a loop, and $|I - J|$ is the distance between the centers of leg I and leg J. ℓ is a function of the configuration of the 3×3 block, $\ell = \ell(C)$; there are $2^9 = 512$ configurations.

The partition function for the 3×3 block is

$$Z = \sum_C \exp\left(-\beta\left(E_1(C) + \ell(C)\mu\right)\right);$$

the probability of a configuration C is $P(C) = Z^{-1} \exp\left(-\beta(E_1 + \ell\mu)\right)$. Cover the plane with 3×3 blocks independent of each other, the structure of each block being chosen with the probability $P(C)$. Search for the values of μ and β for which percolation occurs. Clearly, if $\beta = 0$ all configurations are equally likely and one is back at the $p = 1/2$ percolation threshold of the

---

[13]See, e.g., Grimmett, *Percolation*, 1979.

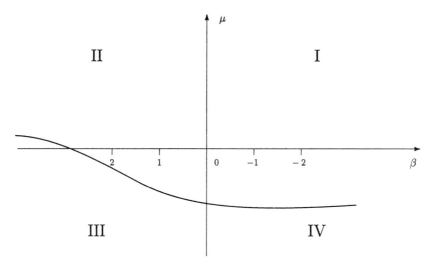

FIGURE 7.8. Percolation loci in (β, μ) plane.

independent plaquettes. However, the correlations introduced by the blocks when $\beta \neq 0$ create an additional curve along which percolation occurs[14]; see Figure 7.8. The fact that the critical states lie along curves is analogous to the presence of a critical line in the two-dimensional Kosterlitz-Thouless renormalization.

The fractal dimension of the long macroscopic vortices remains $7/4$ along both curves.[15] The percolation curves divide the (β, μ) plane into four regions, I, II, III, IV. The area far down where $\mu < 0$ is presumably unphysical; region I corresponds to smooth flow, II to disorderly flow; III can be identified with a superfluid region. One can thus produce a percolation threshold, and with it either turbulence or a transition from a superfluid to a normal state, by either changing T or by changing the cost of producing new vortex legs. The model at hand is clearly too simple, but it creates a rational expectation that the statistics of vortices in turbulence and at the λ transition may be similar and that both can be related to a percolation model. There is evidence[16] that at the λ transition $\mu \cong 0.6$, as is the case when $|T| = \infty$.

This may be the place to emphasize the differences between quantum superfluid vortex motion and classical Euler/Navier-Stokes vortex motion. Quantum vortices are quantized (i.e., Γ can take on only one of a discrete collection of values). More importantly, quantum vortices interact strongly with the fluid in which they are imbedded and take on its posi-

---

[14]A. Chorin, 1992.
[15]H. Saleur and B. Duplantier, 1985; A. Chorin, 1992.
[16]S. Shenoy, 1989.

tive temperature; quantum effects appear when that temperature is small; classical vortices make their own temperature, which is typically negative. Quantum vortices are nearly true lines; classical vortices have a non-trivial cross-section. One does not expect a Kolmogorov law in quantum vortices, and if there is an inertial range, there is no reason to expect the 5/3 exponent; the observed γ should be smaller than 5/3. The possible commonality of classical and quantum critical states does not necessarily imply a commonality in their dynamics. In particular, the constant temperature of quantum vortex states should inhibit vortex stretching and indeed implies that the motion of the vortices cannot be described by a non-singular perturbation of Euler's equations, contrary to what is often stated. In particular, the interaction of superfluid vortices with "phonons", by keeping T fixed, allows the formation of equilibria at finite T with finite vortex length per unit volume.

Finally, we make a brief but important remark: percolation models make possible the construction of relatively dense ensembles of vortex filaments, in which the conclusions of models based on sparse vortex suspensions can be checked numerically.

7.6. Renormalization of Vortex Dynamics in a Turbulent Regime

We have seen that the evolution of a vortex filament towards a turbulent $|T| = \infty$ state resembles the inverse of a renormalization. The natural way to decrease the complexity of a $|T| = \infty$ vortex tangle is to use the machinery of renormalization to reduce the effective number of vortex filaments.

We have already made plausible the idea that the points $|T| = \infty$, $(\beta = 0)$, μ or λ arbitrary, are fixed points of a renormalization parameter flow. Indeed, in the two-dimensional case, equations (6.12) and (6.13) reduce at $\beta = 0$ to the identities $0 = 0$, $\frac{d\lambda^2}{d\lambda} = 2\lambda$, that do not affect the value $\beta = 0$; a similar phenomenon appears in the corresponding three-dimensional equations that we have not written out. These equations do not yield a rule for renormalizing the vortex strengths and cores, but the correct renormalization is clear: $|T| = \infty$, $\mu = \text{constant}$. Indeed, it has already been pointed out that when $|T| = \infty$, a large vortex loop neither polarizes nor antipolarizes smaller loops, and thus the latter can be removed without affecting the former.

It is quite easy to implement such a removal in the framework of a numerical approximation based on a vortex or Buttke loop representation. Since a large loop with fold is a sum of a smooth large loop and a small loop (Figure 7.9), the removal is also a smoothing. The removal can proceed up to a scale $\lambda_{\max}$ where the initial conditions, boundary conditions and stirring forces have not been forgotten and the $\beta = 0$ equilibrium is not

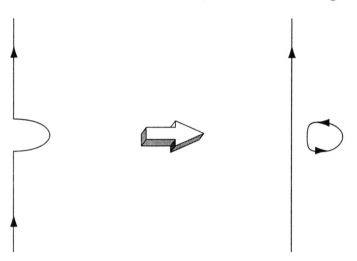

FIGURE 7.9. A folded loop is smoothed.

established. λ_{max} may well be a function of position and time in a non-homogeneous or non-stationary flow. This removal has been shown[17] to reduce computational labor in model problems. It is qualitatively similar, but far from identical, to the usual dealiasing schemes. The following qualitative properties of the removal/renormalization should be noted:

(i) The parameter λ_{max} must either be known in advance or estimated; the renormalization itself does not reveal it.

(ii) The rate of energy dissipation should not be modified by the removal of small-scale vortex structures. With or without such removal, that rate is large in comparison with what would be observed at the same viscosity but in a very smooth flow.

(iii) Much of the small-scale vortical structures having been removed, what remains is a collection of large vortical structures, usually tubular, that are the raw materials for further modeling on scales other than inertial scales (Figure 7.10).

(iv) The effect of vortex removal on small scales is only partially diffusive; nearly parallel vortices are not merged into each other and some small-scale detail remains. It is not in general true that removal can be reasonably modelled by an eddy diffusion coefficient.

(v) The removal of small vortex loops can greatly affect the appearance of large-scale vortex structures, sharpening and tightening the structures that remain.

---

[17] A. Chorin, 1991c; J. Sethian, 1992.

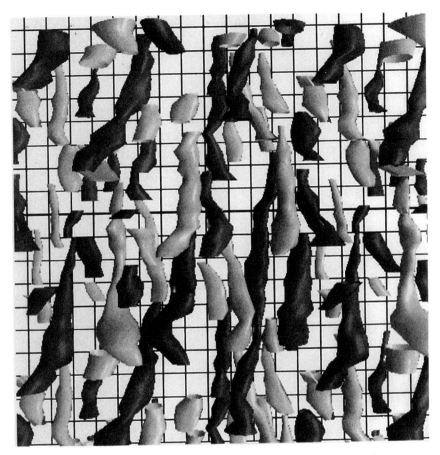

FIGURE 7.10. Large-scale computed smoothed vortices in a boundary region. [Reprinted with permission from P. Bernard, R. Handler, and J. Thomas, *J. Fluid Mechanics*, **253**, 385–419 (1993).]

The problem of implementing a renormalization within the framework of a numerical calculation in a predetermined numerical setting (e.g., a finite difference or spectral method), remains open. The general strategy of large-eddy simulation[18] is sound: one should calculate the large-scale structure from first principles, and remove small-scale structures by a model that takes into account the Kolmogorov scaling that holds at $|T| = \infty$. Large-eddy simulation schemes, by not renormalizing vortex strength, implicitly assume $|T| = \infty$. It is not known at present how to relate the details of the

---

[18]See, e.g., J. Ferziger, 1981; L. Povinelli et al., 1991.

schemes proposed elsewhere to the theory presented here. The renormalization/removal scheme described here cannot be pushed into the realm of non-universal scales above λ_{max} (broader claims have been made on behalf of other renormalization strategies[19]). It would seem that the large scales remain the domain of the inventive modeler and the computational scientist.

The irony in the theory of the inertial range is that it leads, in practice, to methods that allow it to be safely removed from consideration. This is of course a success, because the removal reduces the range of scales that must be resolved in a numerical calculation; such a reduction is a major goal of turbulence theory.

---

[19]See, e.g., D. McComb, 1989; V. Yakhot and S. Orszag, 1986.

Bibliography

It is utterly hopeless to compile a complete bibliography of turbulence theory and of the other topics touched upon in these notes: statistical mechanics, polymers, percolation. The following bibliography is merely a list of the books and papers that I have found useful when I prepared the course on which this book is based. I apologize to all the authors of all the excellent papers and books that are not listed. The omission is not intentional and is not meant as a value judgement.

M. Aizenman, "Geometric analysis of ϕ-4 fields and Ising models", *Commun. Math. Phys.* **86**, 1–48, (1982).

A. Almgren, T. Buttke and P. Colella, "A fast vortex method in three dimensions", in press *J. Comp. Phys.*, 1994.

C. Anderson and C. Greengard, "On vortex methods", *SIAM J. Sc. Stat. Comp.* **22**, 413–440 (1985).

C. Anderson and C. Greengard, *Vortex Methods*, Lecture Notes in Mathematics vol. **1360**, (Springer, New York, 1988).

C. Anderson and C. Greengard, "The vortex ring merger at infinite Reynolds number", *Comm. Pure Appl. Math.* **42**, 1123–1139 (1989).

M. Avellaneda and A. Majda, "An integral representation and bounds on effective diffusivity in passive advection by laminar and turbulent flows", *Comm. Math. Phys.* **138**, 339–391 (1991).

P. Bak and K. Chen, "Self-organized criticality", *Scientific American*, pp. 46–53, (January 1991).

P. Bak, K. Chen and C. Tang, "A forest fire model and some thoughts on turbulence", *Phys. Lett.* **147**, 297–300 (1990).

P. Bak, C. Tang and K. Wiesenfeld, "Self-organized criticality", *Phys. Rev. A* **38**, 364–374 (1988).

G. Batchelor, *The Theory of Homogeneous Turbulence* (Cambridge University Press, Cambridge, 1960).

J. T. Beale, "A convergent 3D vortex method with grid-free stretching", *Math Comp.* **46**, 401–424 (1986).

J. T. Beale, T. Kato and A. Majda, "Remarks on the breakdown of smooth solutions for the 3D Euler equations", *Comm. Math. Phys.* **94**, 61–66 (1984).

J. T. Beale and A. Majda, "Vortex methods I: Convergence in three dimensions", *Math. Comp.* **39**, 1–27 (1982).

J. T. Beale and A. Majda, "Vortex methods II: Higher order accuracy in two and three space dimensions", *Math. Comp.* **32**, 29–52 (1982).

J. T. Beale and A. Majda, "High-order accurate vortex methods with explicit velocity kernels", *J. Comp. Phys.* **58**, 188–208 (1985).

J. Bell and D. Marcus, "Vorticity intensification and the transition to turbulence in the three-dimensional Euler equations", *Comm. Math. Phys.* **147**, 371–394 (1992).

G. Benfatto, P. Picco and M. Pulvirenti, "On the invariant measures for the two-dimensional Euler flow", *J. Stat. Phys.* **46**, 729–742 (1987).

P. Bernard, J. Thomas and R. Handler, "Vortex dynamics in near wall turbulence", *Proceedings International Conference on Near-Wall Turbulence*, C. Speziale and B. Launder, editors (Elsevier, New York, 1993).

P. Bernard, J. Thomas and R. Handler, "Vortex dynamics and the production of Reynolds stress", *J. Fluid Mechanics*, to appear 1994.

A.G. Bershadski, "Large-scale fractal structure in laboratory turbulence, astrophysics, and the ocean", *Usp. Fiz. Nauk.* **160**, 189–194 [*Sov. Phys. Usp.* **33**, 1073–1074 (1990)].

A. Bershadski and A. Tsinober, "Asymptotic fractal and multifractal properties of turbulent dissipative fields", *Phys. Lett. A* **165**, 37–40 (1992).

T. Burkhardt and J. M. J. van Leeuwen (editors), *Real-Space Renormalization* (Springer, Berlin, 1982).

T. Buttke, "Numerical study of superfluid turbulence in the self-induction approximation", *J. Comp. Phys.* **76**, 301–326 (1988).

T. Buttke, "A fast adaptive vortex method for patches of constant vorticity in two dimensions", *J. Comp. Phys.* **89**, 161–186 (1990).

T. Buttke, "Hamiltonian structure for 3D incompressible flow", unpublished manuscript, 1991.

T. Buttke, "Lagrangian numerical methods which preserve the Hamiltonian structure of incompressible fluid flow", in press 1992, *Comm. Pure Appl. Math.*

T. Buttke and A. Chorin, "Turbulence calculations in magnetization variables", to appear in *Proceedings 1992 Israel/U.S. Symposium on Scientific Computing.*

R. Caflisch, *Mathematical Analysis of Vortex Dynamics*, (SIAM, Philadelphia, 1988).

E. Caglioti, P. L. Lions, C. Marchioro and M. Pulvirenti, "A special class of stationary flows for two-dimensional Euler equations: A statistical mechanics description", in press 1991, *Comm. Math. Phys.*

S. Caracciolo, A. Pelissetto and A. Sokal, "Nonlocal Monte-Carlo algorithm for self-avoiding walks with fixed end-points", *J. Stat. Phys.* **60**, 1–53 (1990).

Center for Turbulence Research, Stanford, 1990, Summer school proceedings on studying turbulence using numerical simulation databases.

D. Ceperley and E. Pollock, "Path-integral computation techniques for superfluid ^{4}He, Monte-Carlo methods in theoretical physics", S. Caracciolo and A. Fabrocini, editors (ETS Editirice, Pisa, 1990).

D. Chandler, *Introduction to Modern Statistical Mechanics*, (Oxford, 1987).

C. C. Chang, "Random vortex methods for the Navier-Stokes equations", *J. Comp. Phys.* **76**, 281–300 (1988).

A. Cheer, "Unsteady separated wake behind an impulsively started cylinder", *J. Fluid Mech.* **201**, 485–505 (1989).

A. J. Chorin, "The numerical solution of the Navier-Stokes equations", *Math. Comp.* **22**, 745–762 (1968).

A. J. Chorin, "Inertial range flow and turbulence cascades", Report NYO–1480–135, Courant Institute, New York University, 1969.

A. J. Chorin, "Computational aspects of the turbulence problem", *Proceedings 2nd International Conference on Numerical Methods in Fluid Mechanics* (Springer, New York, 1970).

A. J. Chorin, "Vortex methods for rapid flow", *Proceedings 2nd International Conference on Numerical Methods in Fluid Mechanics* (Springer, New York, 1972).

A. J. Chorin, "Numerical study of slightly viscous flow", *J. Fluid Mech.* **57**, 785–796 (1973).

A. J. Chorin, "Gaussian fields and random flow", *J. Fluid Mech.* **63**, 21–32 (1974).

A. J. Chorin, 1975 *Lectures on Turbulence Theory*, Mathematics Dept. Lecture Notes, University of California, Berkeley.

A. J. Chorin, "Vortex models and boundary layer instability", *SIAM J. Sc. Stat. Comp.* **1**, 1–21 (1980).

A. J. Chorin, "Estimates of intermittency, spectra and blow-up in fully developed turbulence", *Comm. Pure Appl. Math.* **34**, 853–866 (1981).

A. J. Chorin, "The evolution of a turbulent vortex", *Comm. Math. Phys.* **83**, 517–535 (1982).

A. J. Chorin, "Turbulence and vortex stretching on a lattice", *Comm. Pure Appl. Math.* **39** (special issue), S47–S65 (1986).

A. J. Chorin, "Lattice vortex models and turbulence theory", in *Wave Motion*, Lax 60th birthday volume, A. Chorin and A. Majda, editors (MSRI-Springer, 1987).

A. J. Chorin, "Scaling laws in the lattice vortex model of turbulence", *Comm. Math. Phys.* **114**, 167–176 (1988a).

A. J. Chorin, "Spectrum, dimension and polymer analogies in fluid turbulence", *Phys. Rev. Lett.* **60**, 1947–1949 (1988b).

A. J. Chorin, *Computational Fluid Mechanics*, selected papers (Academic, New York, 1989).

A. J. Chorin, "Constrained random walks and vortex filaments in turbulence theory", *Comm. Math. Phys.* **132**, 519–536 (1990a).

A. J. Chorin, "Hairpin removal in vortex interactions", *J. Comp. Phys.* **91**, 1–21 (1990b).

A. J. Chorin, "Vortices, turbulence and statistical mechanics", in *Vortex Methods and Vortex Motions*, K. Gustafson and J. Sethian, editors (SIAM, Philadelphia, 1991a).

A. J. Chorin, "Statistical mechanics and vortex motion", *AMS Lectures in Applied Mathematics* **28**, 85–101 (1991b).

A. J. Chorin, "Hairpin removal in vortex interactions II", report LBL-30927, Lawrence Berkeley Laboratory, 1991c.

A.J. Chorin, "Equilibrium statistics of a vortex filament with applications", *Comm. Math. Phys.* **141**, 619–631 (1991d).

A. J. Chorin, "A vortex model with turbulent and superfluid percolation", *J. Stat. Phys.* **69**, 67–78 (1992).

A. J. Chorin and J. Akao, "Vortex equilibria in turbulence theory and quantum analogues", *Physica D* **52**, 403–414 (1991).

A. J. Chorin and P. Bernard, "Discretization of a vortex sheet with an example of roll-up", *J. Comp. Phys.* **13**, 423–429 (1973).

A. J. Chorin and J. Marsden, *A Mathematical Introduction to Fluid Mechanics* (Springer, New York, 1993).

W. J. Cocke, "Turbulent hydrodynamic line stretching: Consequences of isotropy", *Phys. Fluids* **12**, 2488–2492 (1969).

A. Coniglio, N. Jan, I. Majid and H. E. Stanley, "Conformation of a polymer chain at the theta prime point: Connection to the external parameter of a percolation cluster", *Phys. Rev. B* **35**, 3617–3620 (1987).

P. Constantin, P. D. Lax and A. Majda, "A simple one-dimensional model for the three-dimensional vorticity equation", *Comm. Pure Appl. Math.* **38**, 715–724 (1985).

G. H. Cottet, "Large time behavior for deterministic particle approximations to the Navier-Stokes equations", *Math. Comp.* **56**, 45–60 (1991).

R. J. Donnelly, *Quantized Vortices in Helium* II (Cambridge University Press, 1991).

D. Dritschel, "Contour surgery", *J. Comp. Phys.* **77**, 240–266 (1988a).

D. Dritschel, "The repeated filamentation of two-dimensional vorticity interfaces", *J. Fluid Mech.* **194**, 511–547 (1988b).

H. Dym and H. McKean, *Fourier Series and Integrals* (Academic, New York, 1972).

J. Epiney, "3D *XY* model near criticality", Diploma thesis, E.T.H., Zurich, 1990.

R. Esposito and M. Pulvirenti, "Three-dimensional stochastic vortex flows", preprint, 1987.

D. Evans and G. Morriss, *Statistical Mechanics of Non-Equilibrium Liquids* (Academic, New York, 1990).

G. L. Eyink and H. Spohn, "Negative temperature states and equivalence of ensembles for the vortex model of a two-dimensional ideal fluid", manuscript, 1991.

J. W. Essam, "Percolation theory", *Prog. Phys.* **43**, 833–912 (1980).

J. H. Ferziger, "Higher-level simulations of turbulent flows", NASA report TF-16, Stanford, 1981.

R. Feynman and M. Cohen, "The character of the roton state in liquid helium", *Prog. Theoret. Phys.* **14**, 261–263 (1955).

D. Fishelov, "Vortex methods for slightly viscous three-dimensional flows", *SIAM J. Sci. Stat. Comp.* **11**, 399–424 (1990).

M. Freedman, Z. X. He and Z. Wang, "On the energy of knots and unknots", manuscript, Math Dept., University of California, San Diego, 1992.

U. Frisch, P. L. Sulem and M. Nelkin, "A simple dynamical model of intermittent fully developed turbulence", *J. Fluid Mech.* **87**, 719–736 (1978).

U. Frisch and G. Parisi, "Fully developed turbulence and intermittency", in *Turbulence and Predictability in Geophysical Fluid Dynamics*, M. Ghil et al., editors, pp. 71–88 (North Holland, Amsterdam, 1985).

J. Frohlich and D. Ruelle, "Statistical mechanics of vortices in an inviscid two-dimensional fluid", *Comm. Math. Phys.* **87**, 1–36 (1982).

J. Frohlich and T. Spencer, "The Kosterlitz-Thouless transition in two-dimensional Abelian spin systems and the Coulomb gas", *Comm. Math. Phys.* **81**, 527–602 (1971).

O. Frostman, "Potentiel d'équilibre et théorie des ensembles", thesis, Lund, 1935.

P. G. de Gennes, *Scaling Concepts in Polymer Physics* (Cornell University Press, Ithaca, 1971).

P. G. de Gennes, "Collapse of a polymer chain in poor solvents", *J.*

Physique **36**, L55–L57, (1975).

I. I. Gikhman and A. V. Skorokhod, *Introduction to the Theory of Random Processes* (Saunders, 1969).

J. Goodman, "The convergence of random vortex methods", *Comm. Pure Appl. Math.* **40**, 189–220 (1987).

R. Grauer and T. Sideris, "Numerical computation of 3D incompressible ideal fluids with swirl", *Phys. Rev. Lett.* **67**, 3511–3514 (1991).

C. Greengard, "Convergence of the vortex filament method", *Math. Comp.* **47**, 387–398 (1986).

C. Greengard and E. Thomann, "Singular vortex systems and weak solutions of the Euler equations", *Phys. Fluids* **31**, 2810–2812 (1988).

G. Grimmett, *Percolation* (Springer, New York, 1989).

O. H. Hald, "Convergence of vortex methods II", *SIAM J. Sc. Stat. Comp.* **16**, 726–755 (1979).

O. H. Hald, "Smoothness properties of the Euler flow map", unpublished manuscript, 1987.

O. H. Hald, "Convergence of vortex methods for Euler's equations III", *SIAM J. Num. Anal.* **24**, 538–582 (1987).

M. Head and P. Bandyopadhyay, "New aspects of boundary layer structure", *J. Fluid Mech.* **107**, 297–338 (1981).

E. Hopf, "Statistical hydromechanics and functional calculus", *J. Rat. Mech. Anal.* **1**, 87–141 (1952).

T. Y. Hou and J. Lowengrub, "Convergence of a point vortex method for the 3D Euler equations", *Comm. Pure Appl. Math.* **43**, 965–981 (1990).

M. B. Isochenko, "Percolation, statistical topography and transport in random media", Report DOE/ET–53088–528 (Institute for Fusion Studies, University of Texas, Austin, 1991).

C. Itzykson and J. M. Drouffe, *Statistical Field Theory* (Cambridge University Press, Cambridge, 1989).

J. D. Jackson, *Classical Electrodynamics* (Wiley, New York, 1974).

G. Joyce and D. Montgomery, "Negative temperature states for the two-dimensional guiding center plasma", *J. Plasma Physics* **10**, 107–121 (1973).

J. P. Kahane and R. Salem, *Ensembles Parfaits et Séries Trigonometriques* (Hermann, Paris, 1963).

P. Kailasnath, A. Migdal, K. Sreenivasan, V. Yakhot and L. Zubair, "The 4/5 Kolomogorov law and the odd-order moments of velocity differences in turbulence", unpublished manuscript, 1992.

T. Kambe and T. Takao, "Motion of distorted vortex rings", *J. Phys. Soc. Japan* **31**, 591–599 (1971).

J. Katzenelson, "Computational structure of the N-body problem", *SIAM J. Sc. Stat. Comp.* **10**, 787–815 (1989).

R. Kerr and F. Hussain, "Simulation of vortex reconnection", *Physica D* **37**, 474–484 (1989).

M. Kiessling, "Statistical mechanics of classical particles with logarithmic interactions", *Comm. Pure Appl. Math.*, **46**, 27–56 (1993).

J. Kim and P. Moin, "The structure of the vorticity field in turbulent channel flow", *J. Fluid Mech.* **162**, 339–361 (1986).

M. Kiya and H. Ishii, "Vortex interaction and Kolmogorov spectrum", *Fluid Dynamics Res.* **8**, 73–83 (1991).

R. Klein and A. Majda, "Self-stretching of a pertubed vortex filament I: The asymptotic equation for for deviations from a straight line", *Physica D* **49**, 323–352 (1991a).

R. Klein and A. Majda, "Self-stretching of perturbed vortex filaments II: Structure of solutions", *Physica D* **53**, 267–294 (1991b).

O. Knio and A. Ghoniem, "Three-dimensional vortex methods", *J. Comp. Phys.* **86**, 75–106 (1990).

G. Kohring and R. Schrock, "Properties of generalized 3D $O(2)$ model with suppression/enhancement of vortex strings", *Nuclear Physics B* **288**, 397–418 (1987).

A. N. Kolmogorov, "Local structure of turbulence in an incompressible fluid at a very high Reynolds number", *Dokl. Akad. Nauk SSSR*, **30**, 299–302 (1941).

J. Kosterlitz, "The critical properties of the two-dimensional xy model", *J. Phys. C: Solid State Phys.* **7**, 1046–1060 (1974).

J. Kosterlitz and D. J. Thouless, "Order, metastability and phase transitions in two-dimensional systems", *J. Phys. C: Solid State Phys.* **6**, 1181–1203 (1973).

R. Kraichnan, "Inertial ranges in two-dimensional turbulence", *Phys. Fluids* **10**, 1417–1423 (1967).

R. Kraichnan, "On Kolmogorov's inertial range theories", *J. Fluid Mech.* **62**, 305–355 (1974).

R. Kraichnan and D. Montgomery, "Two-dimensional turbulence", *Rep. Prog. Phys.* **43**, 547–619 (1980).

R. Krasny, "Computing vortex sheet motion", *Proceedings International Congress of Mathematicians*, Kyoto, 1990.

O. A. Ladyzhenskaya, *Mathematical Problems in the Dynamics of Viscous Incompressible Flow* (Gordon and Breach, New York, 1963).

M. Lal, "Monte-Carlo computer simulations of chain molecules", *Molecular Phys.* **17**, 57–64 (1969).

H. Lamb, *Hydrodynamics*, (Dover, New York, 1932).

J. Lamperti, *Probability*, (Benjamin, New York, 1966).

L. Landau and E. Lifshitz, *Statistical Physics*, 3rd edition, part 1, (Pergamon, New York, 1980).

P. Lax, *Hyperbolic Systems of Conservation Laws and the Mathematical Theory of Shock Waves* (SIAM Publications, Philadelphia, 1972).

S. Leibovich and J. Randall, "Solitary waves on concentrated vortices",

J. Fluid Mech. **51**, 625–635 (1972).

M. Lesieur, *Turbulence in Fluids* (Kluwer, Dordrecht, 1990a).

M. Lesieur, "Turbulence et structures coherentes dans les fluides, Nonlinear partial differential equations and their applications", College de France seminar 1989-1990, H. Brezis and J.L.Lions, editors (Pitman, Paris, 1990b).

A. Leonard, "Computing three-dimensional vortex flows with vortex filaments", *Ann. Rev. Fluid Mech.* **17**, 523–559 (1985).

D. G. Long, "Convergence of the random vortex method in two dimensions", *J. Amer. Math. Soc.* **1**, 779–804 (1988).

D. G. Long, "Convergence of random vortex methods in three dimensions", in press, *Math. Comp.*, 1992.

H. Lugt, *Vortex Flows in Nature and Technology* (Wiley, New York, 1983), p.123.

T. Lundgren and Y. Pointin, "Statistical mechanics of two-dimensional vortices", *J. Stat. Phys.* **17**, 323–355 (1977a).

T. Lundgren and Y. Pointin, "Non-Gaussian probability distributions for a vortex fluid", *Phys. Fluids* **20**, 356–363 (1977b).

S. K. Ma, *Modern Theory of Critical Phenomena*, (Benjamin, Boston, 1976).

N. Madras and A. Sokal, "The pivot algorithm: a highly efficient Monte-Carlo method for self-avoiding walks", *J. Stat. Phys.* **50**, 109–186 (1988).

A. Majda, "Vorticity and the mathematical theory of incompressible fluid flow", *Comm. Pure Appl. Math.* **39**, S187–S179, (1986).

A. Majda, "Vorticity, turbulence and acoustics in fluid flow", *SIAM Review* **33**, 349–388 (1991).

A. Majda, "The interaction of nonlinear analysis and modern applied mathematics", *Proceedings International Congress of Mathematicians*, Kyoto, 1990.

B. Mandelbrot, "Intermittent turbulence and fractal dimension: kurtosis and the spectral exponent $5/3 + B$", in *Turbulence and the Navier-Stokes Equations*, R. Temam, editor (Springer, New York, 1976).

B. Mandelbrot, "Fractals and turbulence", in *Turbulence Seminar*, P. Bernard and T. Ratiu, editors (Springer, New York, 1977).

B. Mandelbrot, *Fractals, Form, Chance and Dimension* (Freeman, San Francisco, 1977).

B. Mandelbrot, *The Fractal Geometry of Nature* (Freeman, San Francisco, 1982).

C. Marchioro and M. Pulvirenti, *Vortex Methods in Two-Dimensional Fluid Mechanics* (Springer, New York, 1984).

D. Marcus and J. Bell, "Numerical simulation of a viscous vortex ring interaction with a density interface", preprint UCRL-JC-105183, Livermore National Laboratory, 1991.

P. Marcus, "Numerical simulation of Jupiter's great red spot", *Nature*

331, 693–696 (1988).

P. Marcus, "Vortex dynamics in a shearing zonal flow", *J. Fluid Mech.* **215**, 393–430 (1990).

J. Marsden and A. Weinstein, "Coadjoint orbits, vortices and Clebsch variables for incompressible fluids", *Physica* **7**, 305–323 (1983).

T. Maxworthy, "Turbulent vortex rings", *J. Fluid Mech.* **64**, 227–239 (1974).

T. Maxworthy, "Some experimental studies of vortex rings", *J. Fluid Mech.* **81**, 465–495 (1977).

W. D. McComb, *Physical Theories of Turbulence* (Cambridge University Press, Cambridge, 1989).

J. McWilliams, "The emergence of isolated coherent vortices in turbulent flow", *J. Fluid Mech.* **146**, 21–46 (1984).

H. Meirovitch, "A Monte-Carlo study of the entropy, pressure and critical behavior of the hard-sphere gas", *J. Stat. Phys.* **30**, 681–698 (1984).

C. Meneveau, "Dual spectra and mixed energy cascade of turbulence in the wavelet representation", *Phys. Rev. Lett.* **66**, 1450–1453 (1991).

C. Meneveau and K. R. Sreenivasan, "The multifractal spectrum of the dissipation field in turbulent flow", *Nuclear Physics B* (Proc. Suppl) **2**, 49–76 (1987).

C. Meneveau and K. R. Sreenivasan, "Interface dimension in intermittent turbulence", *Phy. Rev. A* **41**, 2246–2248 (1990).

J. Miller, "Statistical mechanics of Euler equations in two dimensions", *Phys. Rev. Lett.* **65**, 2137–2140 (1990).

J.D. Miller, "Direction-direction correlations of oriented polymers", *J. Stat. Phys.* **63** (1991).

S. Miyashita, H. Nishimori, A. Kuroda and M. Suzuki, "Monte-Carlo simulation and static and dynamic behavior of the plane rotator model", *Prog. Theoret.Phys.* **60**, 1669–1685 (1978).

H. K. Moffatt, "On the degree of knottedness of tangled vortex lines", *J. Fluid Mech.* **35**, 117–132 (1969).

P. Moin, A Leonrad and J. Kim, "Evolution of a curved vortex filament into a vortex ring", *Phys. Fluids* **29**, 955–963 (1986).

D. Montgomery and G. Joyce, "Statistical mechanics of negative temperature states", *Phys. Fluids* **17**, 1139–1145 (1974).

D. Montgomery, W. Matthaeus, W. Stribling, D. Martinez and S. Oughton, "Relaxation in two dimensions and the 'Sinh-Poisson' equation", *Phys. Fluids A* **4**, 3–6, (1992).

D. Montgomery and L. Phillips, "Minimum dissipation and maximum entropy", in *Maximum Entropy and Bayesian Methods*, P. F. Fougere, editor (Kluwer, Dordrecht, 1990).

S. Nemirovski, J. Pakleza and W. Poppe, "Stochastic behavior of vortex filaments", LIMSI/CNRS report 91-14, Orsay, 1991.

L. Onsager, "Statistical hydrodynamics", *Nuovo Cimento*, suppl. to vol. **6**, 279–287 (1949).

S. Orszag, Comments on "Turbulent hydrodynamic line stretching..." *Phys. Fluids* **13**, 2203–2204 (1970a).

S. Orszag, "Analytical theories of turbulence", *J. Fluid Mech.* **41**, 363–386 (1970b).

V. I. Oseledets, "On a new way of writing the Navier-Stokes equations: The Hamiltonian formalism", Comm. Moscow Math. Soc. 1988, translated in *Russ. Math. Surveys* **44**, 210–211 (1989).

G. Papanicolaou, D. Stroock and S. R. S. Varadhan, "A Martingale approach to some limit theorems", in *Statistical Mechanics and Dynamical Systems*, Duke turbulence conference, D. Ruelle, editor, Duke University Series, vol. **3**, 1977.

M. Perlman, "On the accuracy of vortex methods", *J. Comp. Phys.* **59**, 200–223 (1985).

L. A. Povinelli, W. W. Liou, A. Shabbir and T.H. Shih, Workshop on engineering turbulence modelling, NASA Lewis Research Center, 1991.

R. Prasad and K. Sreenivasan, "The measurement and interpretation of fractal dimensions of the scalar interface in turbulent flow", *Phys. Fluids A* **2**, 792–807 (1990).

E.G. Puckett, "A review of vortex methods", in *Incompressible Computational Fluid Mechanics*, R. Nicolaides and M. Ginzburger, editors (Cambridge University Press, 1992).

D. Pullin, P. Jacobs, R. Grimshaw and P. Saffman, "Instability and filamentation in finite-amplitude waves on vortex layers of finite thickness", *J. Fluid Mech.* **209**, 359–384 (1989).

A. Pumir and R. M. Kerr, "Numerical simulation of interacting vortex tubes", *Phys. Rev. Lett.* **58**, 1636–1639 (1987).

A. Qi, "Three-dimensional vortex methods for the analysis of propagation on vortex filaments", Ph.D. thesis, Math Dept., University of California, Berkeley, 1991.

D. Revuz and M. Yor, *Continuous Martingale Calculus*, Chapter 13 (Springer, New York, 1989).

R. Robert, "A maximum entropy principle for two-dimensional perfect fluid dynamics", *J. Stat. Phys.* **65**, 531–554 (1991).

P. H. Roberts, "A Hamiltonian theory for weakly interacting vortices", *Mathematica* **19**, 169–179 (1972).

S. Roberts, "Convergence of random walk methods", Ph.D. thesis, University of California, Berkeley, 1986.

C. A. Rogers, *Hausdorff Measures* (Cambridge University Press, 1970).

M. Rogers and P. Moin, "The structure of the vorticity field in homogeneous turbulent flows", *J. Fluid Mech.* **176**, 33–66 (1987).

A. Rouhi, "Poisson brackets for point dipole dynamics in three dimensions", unpublished manuscript, University of California, San Diego, 1990.

H. Saleur and B. Duplantier, "Exact determination of the percolation hull exponent in two dimensions", *Phy. Rev. Lett.* **58**, 2325–2328 (1987).

R. Savit, "Duality in field theory and statistical systems", *Rev. Modern Physics* **52**, 453–487 (1980).

J. Sethian and A. Ghoniem, "A validation study of vortex methods", *J. Comp. Phys.* **74**, 283–317 (1988).

K. Shariff and A. Leonard, "Vortex rings", *Ann. Rev. Fluid Mech.* **24**, 235–279 (1992).

Z. S. She, E. Jackson and S. Orszag, "Intermittent vortex structures in homogeneous isotropic flow", *Nature* **344**, 226–228 (1990).

Z. S. She, "Intermittency and non-gaussian statistics in turbulence", *Fluid Dynamics Research* **8**, 143–158 (1991).

H. Shen, "Stochastic mechanics approach to turbulent hairpin evolution", *Physica D*, **51**, 555–566 (1991).

S. R. Shenoy, "Vortex loop scaling in the three-dimensional XY ferromagnet", *Phys. Rev. B* **40**, 5056–5068 (1989).

E. Siggia and A. Pumir, "Incipient singularity in the Navier-Stokes equations", *Phys. Rev. Lett.* **55**, 1749–1750 (1985).

A. Sommerfeld, *Thermodynamics and Statistical Mechanics* (Academic, New York, 1964).

C. Speziale and P. Bernard, "The energy decay in self-preserving istropic turbulence revisited", *J. Fluid Mech.* **241**, 645–667 (1992).

K. R. Sreenivasan, "A unified view of the origin and morphology of turbulent boundary layer structure", IUTAM Symposium, Bangalore, H. W. Liepmann and R. Narasimha, editors, 1987.

K.R. Sreenivasan, "Fractals and multifractals in fluid turbulence", *Ann. Rev. Fluid Mech.*, in press.

K. R. Sreenivasan and C. Meneveau, "The fractal aspects of turbulence", *J. Fluid Mech.* **173**, 357–386 (1986).

D. Stauffer, "Scaling properties of percolation clusters", in *Disordered Systems and Localization*, C. Castellani, C. diCastro and L. Peliti, editors, Springer Lecture Notes in Physics **149** (Springer, New York, 1981).

D. Stauffer, *Introduction to Percolation Theory* (Taylor and Francis, London, 1985).

D. Summers, "An algorithm for vortex loop generation", Lawrence Berkeley Laboratory report LBL-31367, Berkeley, CA, 1991.

H. Tennekes, "Simple model for the small scale structure of turbulence", *Phys. Fluids* **11**, 669–670 (1968).

R. Temam, *The Navier-Stokes Equations* (Elsevier, Amsterdam, 1984).

C. Thompson, *Classical Equilibrium Statistical Mechanics* (Clarendon, Oxford, 1988).

S. Widnall, "The structure and dynamics of vortex filaments", *Ann.*

Rev. Fluid Mech. **8**, 141–165 (1976).

F. W. Wiegel, *Introduction to Path Integral Methods in Physics and Polymer Science* (World Scientific, Singapore, 1986).

R. Van Buskird and P. Marcus, "Vortex dynamics in flows with non-uniform shear", submitted for publication, 1992.

M. van Dyke, *An Album of Fluid Motion* (Parabolic Press, Stanford, 1982), pp. 66–68.

F. Varosi, T. Antonsen, Jr. and E. Ott, "The spectrum of fractal dimensions of passively convected scalar gradients in chaotic fluid flows", *Phys. Fluids A* **3**, 1017–1028 (1991).

A. Weinrib and S. Trugman, "A new kinetic walk and percolation perimeters", *Phys. Rev. B* **31**, 2993–2997 (1985).

F. W. Wiegel, *Introduction to Path Integral Methods in Physics and Polymer Science* (World Scientific, Singapore, 1986).

S. Wiggins, *Introduction to Applied Dynamical Systems and Chaos* (Springer, New York, 1990).

G. Williams, "Vortex ring model of the superfluid lambda transition", *Phys. Rev. Lett.* **59**, 1926–1929 (1987).

G. Williams, "Vortices and the superfluid ^{4}He transition in two and three dimensions", in *Excitations in Two-Dimensional and Three-Dimensional Quantum Fluids*, A. G. F Wyatt and H. J. Lauter, editors (Plenum, New York, 1991).

G. Williams, "Vortex rings and finite wave-number superfluidity near the ^{4}He lambda transition", *Phys. Rev. Lett.* **68**, 2054–2057 (1992).

A. Yaglom, *An Introduction to the Theory of Stationary Random Functions* (Dover, New York, 1962).

V. Yakhot and S. Orszag, "Renormalization group analysis of turbulence I: Basic theory", *J. Sci. Comp.* **1**, 3–51 (1986).

H. Yamakawa, *Modern Theory of Polymer Solutions* (Harper and Row, New York, 1971).

K. Yamamoto and I. Hosakawa, "A decaying isotropic turbulence pursued by the spectral method", *J. Phys. Soc. Japan* **57**, 1532–1535 (1988).

R. Zeitak, "Vectorial correlations on fractals: applications to random walks and turbulence", manuscript, Weizmann Institute of Science, 1992.

Index

169

Applied Mathematical Sciences

(continued from page ii)